PLANNING THE GOOD COM

People either love new urbanism or hate it. Some find compact new neighbourhoods of brownstone row houses, elegant Victorian mansions, or country cottages delightful: places that celebrate the city and its history, and offer hope for a sustainable future. Others see these 'urban villages' as up-graded suburbs mired in the aesthetics of another time and place: cloyingly nostalgic anachronisms for affluent élites. This book examines new urban approaches both in theory and practice. Taking a critical look at how new urbanism lives up to its theory in its practice, it asks whether new urban approaches offer a viable path to the good community.

With examples drawn principally from the United States, Canada, Britain, Germany, Belgium, Norway, and Japan, this book explores new urban approaches in a wide range of settings. It considers the relationship between the movement for urban villages and an urban renaissance that has spread in the UK and Europe with the 'New Urbanism' movement in the United States and Canada and asks whether the concerns that drive contemporary planning theory – issues like power, democracy, spatial patterns, and globalization – receive adequate attention in new urban approaches. Does new urbanism offer a persuasive normative theory of urban development that will shape planning practice for years to come, or a design paradigm that cannot transcend its cultural origins in a particular time and place?

The work of new urbanists has resulted in beautiful urban districts that reveal the potential of planning to create more attractive and meaningful urban landscapes. New urbanists have developed and propagated a formula for planning the good community, and have gained international attention in the process. Beauty is arguably a necessary condition for the good community, but is it sufficient?

Jill Grant is Director of the School of Planning at Dalhousie University, Canada.

THE RTPI Library Series

Editors: Cliff Hague, Heriot Watt University Edinburgh, Scotland
Tim Richardson, Sheffield University, UK
Robert Upton, RTPI, London, UK

Published in conjunction with The Royal Town Planning Institute, this series of leading-edge texts looks at all aspects of spatial planning theory and practice from a comparative and international perspective.

Planning in Postmodern Times
Philip Allmendinger, University of Aberdeen, Scotland

The Making of the European Spatial Development Perspective
No Master Plan
Andreas Faludi and Bas Waterhout, University of Nijmegen, The Netherlands

Planning for Crime Prevention
Richard Schneider, University of Florida, USA and Ted Kitchen, Sheffield Hallam University, UK

The Planning Polity
Mark Tewdwr-Jones, The Bartlett, University College London

Shadows of Power
An Allegory of Prudence in Land-Use Planning
Jean Hillier, Curtin University of Technology, Australia

Urban Planning and Cultural Identity
William JV Neill, Queen's University, Belfast

Place Identity, Participation and Planning
Edited by Cliff Hague and Paul Jenkins

Planning for Diversity
Policy and Planning in a World of Difference
Dory Reeves

Planning the Good Community
New Urbanism in Theory and Practice
Jill Grant

Forthcoming:
Indicators for Urban and Regional Planning
Cecilia Wong, University of Liverpool

Planning, Law and Economics
Barrie Needham

Planning at the Landscape Scale
Paul Selman

PLANNING THE GOOD COMMUNITY
NEW URBANISM IN THEORY AND PRACTICE

JILL GRANT

LONDON AND NEW YORK

First published 2006 by Routledge
2 Park Square, Milton Park, Abingdon, Oxon OX14 4RN

Simultaneously published in the USA and Canada
by Routledge
270 Madison Ave, New York, NY 10016, USA

Routledge is an imprint of the Taylor & Francis Group

© 2006 Jill Grant

Transferred to Digital Printing 2007

Typeset in Akzidenz Grotesk by RefineCatch Limited, Bungay, Suffolk
Printed and bound in Great Britain by Cpod, Trowbridge, Wiltshire

All rights reserved. No part of this book may be reprinted or reproduced or utilised in any form or
by any electronic, mechanical or other means, now known or hereafter invented, including photocopying
and recording, or in any information storage or retrieval system, without permission in writing
from the publishers.

The publisher makes no representation, express or implied, with regard to the accuracy
of the information contained in this book and cannot accept any legal responsibility or liability for
any errors or omissions that may be made.

British Library Cataloguing in Publication Data
A catalogue record for this book is available from the British Library

Library of Congress Cataloging in Publication Data
A catalog record for this book has been requested

ISBN10: 0–415–70074–4 (Hbk)
ISBN10: 0–415–70075–2 (Pbk)

ISBN13: 978–0–415–70074–0 (Hbk)
ISBN13: 978–0–415–70075–7 (Pbk)

To my students – past, present and future – who challenge and inspire me in so many ways.

CONTENTS

List of figures — VIII
List of tables — X
List of boxes — XI
Illustration credits — XII
Notes on contributors — XIII
Preface — XV
Acknowledgements — XXII
List of terms and measurements — XXIV

PART 1
THE RISE OF NEW URBAN APPROACHES

01	New urbanism(s) ascendant	3
02	Getting to (sub)urbanism	29
03	Theory in new urbanism	45

PART 2
NEW URBANISM(S) IN PRACTICE

04	New urbanism is born: the American experience	81
05	Revamping urbanism: the European experience	105
06	Modernizing urbanism: new urban Asia	131
07	Colonial urbanism: Canada signs on	151

PART 3
THE PROSPECTS FOR NEW URBANISM(S)

| 08 | Reconciling new urbanism's theory and practice | 175 |
| 09 | The fate of new urbanism | 203 |

Notes — 231
Bibliography — 235
Index — 265

LIST OF FIGURES

1.1	Handsome civic spaces	5
2.1	The garden city reduced	39
3.1	A traditional street in Nordnes	48
3.2	The European city square	53
3.3	The transit station as community hub	59
3.4	Villages for the Millennium	61
3.5	Transect/concentric zone theory	75
4.1	Seaside site plan	83
4.2	Kentlands Main Street	87
4.3	Kentlands townhouses	87
4.4	King Farm shuttle bus	89
4.5	King Farm shops	90
4.6	Fallsgrove 'village center' mall	90
4.7	Orenco Station house	91
4.8	Santana Row: the mall goes retro	93
5.1	The medieval city as the good community	107
5.2	Upton Village site plan	115
5.3	Rue de Laeken: new traditions	116
5.4	Karow Nord: modernist new urbanism	117
5.5	Poundbury: classical market	121
5.6	Greenwich Millennium Village: Legoland apartments	121
6.1	No 'eyes on the street'	137
6.2	Winding alleys lead home	138
6.3	Eclectic mix of styles in Tokyo	140
6.4	White Town site plan	142
6.5	White Town street: a Western feel	143
6.6	Hikone: new traditional urbanism	147
7.1	Don Mills: garden city Canadian style	153
7.2	St Lawrence mixed use	154
7.3	McKenzie Towne site plan	158
7.4	McKenzie Towne gazebo	159
7.5	Cornell: upscale country	162
7.6	Cornell town centre	163
7.7	Garrison Woods: urban rehab	165

7.8	Skinny house: an excuse for ugliness?	169
8.1	New urban redux	179
8.2	Whose history? Northern cottages	182
8.3	Fantasy Village	193
9.1	Lakelands street	208
9.2	McKenzie Towne High Street	209

LIST OF TABLES

1.1	The key values of planning in planning movements	21
1.2	Sample vision statements	25
2.1	Models of good community form	32
3.1	Comparing the principles of new urban approaches	57
5.1	European organizations promoting traditional urbanism/new urban approaches	113
7.1	Comparative statistics on McKenzie Towne and McKenzie Lake	160
9.1	Selling community	205

LIST OF BOXES

3.1	Strong traditional neighbourhoods: Nordnes *Todd Saunders*	48
4.1	Kentlands: American dreams and realities *Patrick Moan*	85
4.2	Santana Row: the mall becomes a village *Sue Beazley*	93
5.1	Karow Nord: rebuilding the German village *Trevor Sears*	117
5.2	Poundbury: not quite real *Marc Ouellet*	122
6.1	White Town: American-style suburbanism *Jill Grant and Kaori Itoh*	142
6.2	Hikone rediscovers its history *Jill Grant*	147
7.1	McKenzie Towne: urbanism in the Prairies *Jill Grant*	159
7.2	Cornell: new town on the leading edge *Jill Grant*	162

ILLUSTRATION CREDITS

Sue Beazley 4.8
Marc Ouellet 5.5
Matt Reid 2.1, 3.5, 4.1, 5.2, 6.4, 7.3, 8.3
Todd Saunders 3.1
Trevor Sears 5.4

NOTES ON CONTRIBUTORS

Through the years many students and research assistants have helped me with my work and inspired me to ask and answer challenging questions. I invited several of them to contribute their thoughts on some of the communities they have studied and worked on in recent years. Their musings appear in text boxes in several chapters. I appreciate their participation in the discussion.

Sue Beazley studied at Carleton and the Nova Scotia College of Art and Design. A Project Planner with Design, Community and Environment in Berkeley, her work focuses on environmental review under the California Environmental Quality Act. She regularly contributes to developing general plans for San Francisco Bay area jurisdictions. Reach her at sue@dceplanning.com

Jill Grant is Professor of Planning at Dalhousie University in Halifax, Canada. She authored *The Drama of Democracy: Contention and Dispute in Community Planning* (University of Toronto Press), and co-edited *Towards Sustainable Cities: East Asian, European, and North American Perspectives on Managing Urban Change* (Ashgate). Contact her at jill.grant@dal.ca

Kaori Itoh graduated from the Graduate School of International Development and is currently working on her doctorate at Nagoya University in Japan: her dissertation research examines women's economic development activities in Meiho village. She works for the Gifu Prefectural government, and can be contacted at kao@va.u-netsurf.jp

Patrick Moan has a Masters degree in Planning from Dalhousie University. In Baltimore County, Maryland, he helped integrate new urbanist design charrettes into a redevelopment policy targeting distressed inner-ring suburbs. He works with a systems engineering firm in Washington DC on modernizing the American air traffic management and security systems. Reach him at pcmoan@earthlink.net

Marc Ouellet completed degrees from the Université de Moncton, Concordia University, and Dalhousie University. His Masters thesis examined connections between major designers in American and European new urbanism. He currently works as a planner for the Beaubassin Planning Commission in New Brunswick. He can be reached at marc@beaubassin.nb.ca

Matt Reid graduated with an honours Bachelor of Community Design degree with a Major in Urban Design Studies from Dalhousie University, Nova Scotia, Canada

(2005). He has exhibited his design work and illustrations in group shows at Dalhousie University and plans to teach English in Japan before furthering his education. Contact him at mrreid@dal.ca

Todd Saunders has degrees in Environmental Planning (NSCAD) and Architecture from McGill. Principal of Saunders Architecture in Bergen, Norway, he teaches part-time at the Bergen Architecture School. With awards from the Norwegian Association of Artists and the national government, he was featured in a Norwegian Architecture Association exhibit in 2004. Reach him at post@saunders.no

Trevor Sears, originally from Halifax, Canada, has a Masters in Landscape Architecture (University of Guelph), and a Bachelor of Design in Environmental Planning (Nova Scotia College of Art and Design). Since 1996 he has worked for landscape architecture offices in Berlin and Amsterdam. He currently resides in Berlin where he can be reached at trevor.sears@freenet.de

PREFACE

People either love new urbanism or hate it. Some find compact new neighbourhoods of brownstones, nineteenth-century style houses, or country cottages delightful: places that celebrate the city and offer hope for a sustainable future. Others see these places as upgraded suburbs mired in the aesthetics of another time and place: cloyingly nostalgic anachronisms. Can we bridge this divide to see the merit in both positions? I intend to try.

When I first thought about what this book might achieve, I had considered writing a comprehensive overview of new urbanism and new urban approaches around the world. As I got more deeply into the project, I realized the hubris in that ambition. The practice is expanding so quickly in recent years that I cannot hope to document it fully. Hence, this is *not* a handbook on new urban projects. I discuss some specific projects, and offer commentaries on others by planners familiar with them. However, I do not present detailed case studies. Those looking for case studies on new urbanist projects can read a growing number of books and journal articles that offer insightful lessons: I leave that task to others. Instead, I present a general overview along with some personal perspectives to consider the ways in which the projects live up to the theory of new urbanism in the realities of their practice.

I take a particular view of theory and practice in this book. I cannot hope to be comprehensive in either theory or practice. The planning theory literature has become so broad and deep that I do not presume to have mastered it by any stretch. My ambition is to stimulate practitioners to consider the theoretical implications of what they are doing and to ask theoreticians to understand the practical appeal of new urbanist ideas. I hope to contribute to the dialogue between theory and practice by exploring the new urbanism and related movements in a wider range of settings than other authors have usually considered.

I recognize and acknowledge the limitations of what I can offer. In two-and-a-half decades, new urbanism has generated a vast literature: more appears daily. In almost a decade of following the movement with keen interest, I still feel I have only skimmed the surface. With hundreds of firms and countless planners employing new urban approaches in a dozen or more countries, a huge range of practice and experience warrants documentation. I cannot do that here. I have had to satisfy myself with raising the key questions, identifying the major themes, and highlighting some interesting examples.

As a planning educator, I am neither a theoretician nor a practitioner. I am merely one of those scholars who tries to understand and illuminate the relationship between theory and practice. This, of course, leaves me vulnerable to criticisms from both camps. Fair enough. I recognize that I can only touch the surface, raise questions, and point to connections. Since that needs doing, I relish the challenge.

I have spent more than two decades studying residential environments and the context of community planning. The places where people choose to live, and the ideas that planners bring to managing those places fascinate me. It's probably the anthropologist in my background that accounts for this curiosity for the mundane.

When I switched from anthropology to planning, I trained as a policy planner. Gradually through the years, though, I found myself gravitating towards studying design questions and issues of practice. I admit to seeing new urbanism as immensely attractive in the physical beauty that it can deliver. When Seaside first gained notoriety, I admired the way in which the principles that inspired it connected to values and history I hold in high regard. Many of the places built in the name of new urbanism proved immediately elegant and captivating. I saw neo-traditional town planning as a promising method for reconciling some of the challenges of the contemporary city. I found hope in its prescriptions for ending wasteful sprawl, providing housing for all, and enhancing the beauty of the public realm. These are noble causes indeed.

Over time, though, I have become wary of the evangelical zeal of the new urbanists, and the cult of celebrity that surrounds new urbanism's exponents. I find vitriol and hyperbole distasteful. Some new urbanists have proven especially prone to *ad hominem* arguments that discredit their own contributions. New urbanism now involves big dollars and huge egos. Some of the simplicity and promise of the concept has been lost in translation and accumulation.

I also suspect notions of universal truths, 'human nature', and 'the public interest'. Concepts that reduce difference with claims of inherent commonality defy my understanding of the real diversity that the world holds. New urbanists follow some powerful thinkers in asserting these universals, but I cannot accept the logic. My background in anthropology makes me keenly aware of the vast range of differences in people and cultures; my years of teaching and studying planning history reveal the significance of context in understanding the choices that people face and make. Were it true that one could find universal reason, that would certainly facilitate human science: I just cannot believe that one can. Indeed, I worry that claims of universality serve in some ways to mask the exercise of authority: the designer as expert imposes his/her vision with the assertion of its universal application. As these illusions became a more significant part of the rhetoric of new urban approaches, I found myself questioning the paradigm.

As someone who educates prospective planners and has worked with and studied planning practitioners, I also admit to discomfort with an ideology that blames planners for the ills of the city without looking at the wider context in which planning occurs. Not that I would claim that planners are mere victims of 'the system', but they are relatively minor players in a big game over which they control few rules. Planning practice is intrinsically embedded within a political economy dependent on urban growth. As a government service, planning provides a cultural mechanism for managing and enabling that growth, along with the pursuit of other cultural values.

I also acknowledge a certain suspicion that new urbanism represents an architectural movement seeking to displace planning. In the urban design arena planners and architects may be rivals as well as colleagues. New urban approaches render policy irrelevant by replacing words with pictures, and goals with graphics. They may privatize the planning process by using invitational charrettes or by substituting private property codes for public rules and zones. I admit to believing in planning, to thinking that we can write good planning policy, and to favouring public processes over private solutions.

In my view, planning must deal with issues of equity, power, and environmental sustainability. Can new urbanism deliver on its promises in these areas? I find it deeply troubling that new urbanism may provide justifications for reducing the number of public housing units in the USA at a time when so many are homeless. I reject the implicit new urban argument that growth can be good if only we get the form correct: this logic ignores the environmental consequences of escalating consumption processes and the social consequences of economic stratification. I am dismayed that new urbanists can suggest that affordable housing is merely desirable, while they insist that upscale housing is essential. For some reason I find myself sceptical of proponents who are fabulously wealthy, and designers who add foreign accents to their names in mid-life. I cannot ignore the class dimensions of the movement. I guess that working class roots make me something of a socialist at heart.

Despite my personal reservations, I have to admire the dissemination of these ideas. New urbanist principles have spread widely, just as garden city notions before them. Projects have even been attempted in some unlikely places. For instance, Melrose Arch in Johannesburg (South Africa) is an upper end urban village concept. It has proven a huge success financially. As Landman (2003) notes, though, one cannot easily compact the apartheid city: its segregation proves too resilient. Efforts to implement urban village principles in this context are likely to take oppressive forms, like gated enclaves. Even in societies without the anguished history of South Africa, security concerns may distort initiatives to restore a vibrant urbanism. We can take planning ideas around the world, but we cannot control the ways in which they or others apply them.

Studying planning theory and practice leads me to argue that we use planning to help us manage problems that are deeply embedded in our societies. We hope that good planning policy and excellent urban design might frame the good community. Sometimes we fail to recognize the economic and power structures within which we operate. We imagine a more powerful role for planning than it can ever achieve in contemporary cultures. In expecting that planning and design can civilize the city, we underestimate the influence of a vast cultural complex that generates ideas of appropriate behaviour. If television commercials selling back-to-school supplies use actors who model 'bad' behaviour to children, and if programmes that promote foul language and graphic violence are popular daily viewing, then how can we believe that building attractive places may generate the good community? Good design cannot cure a sick society. We operate within a popular culture that in many Western nations increasingly advocates bad behaviour. In that context, we might conclude that new urbanism does not present a real critique of our communities or lifestyles, but rather window-dressing to mask the ills of contemporary urban society by building prettier settings.

Although a book on new urbanism is not the place for me to raise issues with contemporary theory in planning, I should flag a few reservations. While I believe that the focus on communication and collaboration in theory is important, I worry that such concerns do not help practitioners decide what kinds of places to build. The issues that have preoccupied theoreticians in recent years have little immediate consequence or relevance for planners. Practitioners understand the need to be truthful and open (and they believe that they are), but how do they know what a planning policy should say about new development, or whether a particular project is good or bad for the city in the long term? Useful theory has to offer guidance to practitioners in their everyday work.

We need to find ways to bridge theory and practice, and to encourage reflective learning. We must develop theory that speaks meaningfully to practitioners and engenders practice that offers new lessons for theory. I hope this book contributes to the debate.

THE ORGANIZATION OF THE BOOK

I have organized the book in three sections. Section I looks at the rise of new urban approaches, setting out the historical and theoretical background that led to the movement. Section II examines new urbanism in practice, developing accounts of the experience with new urban approaches in several countries. Section III considers the prospects for new urbanism and new urban approaches into the future.

In the first chapter I introduce new urbanism and try to understand its rapid rise to prominence in planning. I consider how new urbanism is shaping, responding to, or drawing on planning theory. New urbanism has often been accused of being anti-theoretical, but we see that new urban approaches do use theory, if quite selectively. I suggest that new urbanism has succeeded in effectively framing the popular discussion about how to plan the good community. In that process, it has narrowed the debate and limited consideration of social justice issues.

Chapter 2 begins by briefly exploring what planning theory has traditionally said about urban form and community character. In the early twentieth century, the garden city became the dominant theory of urban form in planning. Through the post-war period, the garden city concept influenced planning practice on almost every continent. Planning and development practice, however, reduced the paradigm to the 'garden suburb' at best, and sprawling formless development at worst; thus it robbed the original garden city model of its theoretical promise. I consider the contributions of the garden city paradigm to understand its wide cross-cultural appeal and application. Garden city ideas strongly influenced theories of physical planning until the late twentieth century, although practice increasingly diverged from the ideal. Dominant theories reflect the broad social and political context in which they take shape, while practice is more firmly rooted in local traditions and concerns. Theories influence practice where they speak to local issues as well as to global concerns.

Chapter 3 introduces the theory of new urbanism and its related manifestations. Although the proponents of new urbanism downplay the significance of theory, the movement operates according to identifiable premises, assumptions, and hypotheses about the effects interventions may have. As the movement is still taking shape, the theories of new urbanism remain multiple, partial, and sometimes internally contradictory. New urban approaches often look to the past for lessons in urban organization and building patterns. As its historicism and spatial determinism has come under attack, however, new urbanism's proponents have increasingly turned to ecological theory for inspiration. Through setting out the spatial conditions for generating good social behaviour and connections, new urbanism tries to frame theory and practice for the good community.

Section II of the book offers analyses of new urban approaches in several parts of the world. Cross-cultural comparisons give us a better understanding of the elements of theory and practice that have the greatest impact and dissemination. Chapter 4 discusses the influence of new urbanism in America. It examines the movement in a social and professional context to explain how it came to displace the garden city paradigm in theory, and often in practice. New urbanism resonated with key values in American culture in a time of fiscal conservatism. When American cities were suffering decay and despair, new urbanism identified a safer and simpler time as its model of the good community.

Chapter 5 looks at how new urbanism and new urban approaches have been received in Europe. With its long legacy of fine cities, compact development, and mixed use, Europe is often cited as offering examples of planning principles that new urban approaches seek to reinforce. In recent years, however, elements of new urbanism have inspired planning approaches and practice in the UK and Europe. How do the European compact city model and the 'urban villages' concept draw on, relate to, extend, or challenge the principles of new urbanism?

East Asian urbanism already meets many of the design principles associated with new urbanism. Development is compact, mixed use, and transit-oriented. Does this mean that new urbanism rules in Asia, or that Asian strategies are more sustainable? Chapter 6 looks at new urban approaches in East Asia, with a special focus on Japan to consider whether new urbanism resonates with the Japanese experience. While we find evidence of cultural borrowing of some ideas, such as planned new towns with American features and commercial precincts in retro-styles, we have to conclude that Japanese urbanism remains essentially modernist.

Closely linked culturally and economically to the USA, Canada often follows American trends. Chapter 7 explains that the leaders of the new urbanist movement found a welcome reception in Canada. Even before the 1980s, the principles that came to be associated with new urbanism were already driving urban infill development in some cities. Several large-scale projects gained considerable publicity within the new urbanist movement in the 1990s, but some experienced economic difficulties in the market. I explore the reasons why Canadian planners have found new urbanism such a persuasive paradigm, and explain some of the challenges to implementation in practice.

Section III considers the implications of practice for the theory of new urbanism. Chapter 8 tries to reconcile new urbanism's theory with its practice. While many practitioners make no bones about believing that they can influence social behaviour through good design, the theorists of the movement deny that environmental determinism is a major premise of new urbanism. I consider the trajectory of new urbanism: how does its practice reveal the limitations of its theory? Are its practices creating places that are affordable, equitable, democratic, diverse, and sustainable?

Chapter 9 discusses the fate of new urbanism. While new urbanism has made a significant impact on urban development practices in recent decades, I predict that its long-term effects may be limited. In practice new urbanism is already being watered-down or cherry-picked. Architectural features like front porches have caught on with developers who recognize useful marketing devices when they see them. Weaning people from their cars and from their large lots is proving more of a challenge. Moreover, practice reveals some of the ironies of new urbanism as a theoretical paradigm.

As planning theorists continue to work towards developing a theory of good city form, some new urbanist ideas may remain influential, while others become historical artifacts. New urbanism may have some longevity, primarily because it provides a powerful marketing device for developers competing to sell homes. Ultimately, though, history will likely reveal the weakness of new urbanism: namely, its commitment to beautiful urban form rather than to social or political reform as the means to create good communities.

A cross-cultural examination suggests that new urbanism's assumptions about the nature of the city reflect particular temporal and spatial politics that may not apply outside of North America. New urbanism offers design solutions with a particular normative approach that is not as widely shared as was the garden city paradigm. It fails to address issues of power and participation. Thus its potential to contribute to wider theoretical debates is reduced. Its claim to have discovered the formula for the good community may be challenged by those who believe that good communities have to be much more than beautiful places.

Since contemporary mainstream planning theory focuses on issues of power and communication, new urbanism is not proving as powerful a theoretical construct as its advocates may hope. Moreover, at the same time as new urbanism resonates with those who embrace a return to urban values, other key development trends in suburban form perpetuate radically different premises. For instance, a much larger proportion of new American developments feature walls and gates than those representing new urbanist principles. Rather than presaging a return to beautiful and urbane landscapes, new urbanism may represent a valiant but vain effort to stave off the urban fragmentation that threatens the contemporary city in the early twenty-first century.

ACKNOWLEDGEMENTS

Ideas for this book have rattled around in my head for several years. In that time, many people have contributed to my thinking and my activities. I thank them all, jointly and severally.

I owe particular gratitude to the research assistants who helped me through various phases of this work through a decade of data collection and thinking about these issues. Thanks especially to Joseph Driscoll, Darrell Joudrey, Jen Meurer, Jaime Orser Smith, and Mark TeKamp.

Funding for the research was supported by three agencies, and my university. Canada Mortgage and Housing Corporation financed the research on sustainable development, and the Social Sciences and Humanities Research Council of Canada provided resources for the research on new urbanism (and more recently on gated communities). The Japan Foundation awarded me a Research Fellowship that allowed me to conduct field studies in Japan. Dalhousie University provided supplemental funds to help cover the costs of a research trip to Europe. These organizations bear no responsibility for the opinions expressed, but without their assistance I would have little evidence to back up my opinions. I thank them for their generous contributions to planning research.

Sue Beazley, Patrick Moan, Marc Ouellet, Todd Saunders, and Trevor Sears generously agreed to turn a critical eye toward urban projects they had studied or visited. I appreciate their willingness to help out their former professor and become partners in the writing. Kaori Itoh, who helped me in countless ways during my sabbatical in Japan, co-authored the vignette on White Town. I thank her, Professor Hisafumi Saito, Machiko Saito, and the faculty and staff at Chukyo University for all of their assistance in facilitating my research on community design in Japan.

Many planners and community residents proved extraordinarily helpful in providing information, agreeing to interviews, and leading tours of some of the places described in the book. These people care deeply about their communities, and put their hearts and souls into their efforts to create good communities. I appreciate the job they do, and thank them for their many kindnesses.

The beautiful hand-drawn illustrations are the work of Matthew Reid, a recent graduate of our planning school. I thank him for giving life to my concepts.

Friends and family also contributed to this project in various ways. Alex Grant, Joyce MacCulloch, and Wendy Cherry provided invaluable logistical support and company on field trips in the UK. Sari Zelenietz helped with organizing the bibliographic material. Christine Bae and Harry Richardson gave me a lift to Kentlands, and Patrick and Niall Moan devoted a day to leading us on a tour of Kentlands and Lakelands. I appreciate the matchmaking of David Gordon who advised me that RTPI was interested in a book on new urbanism, and suggested to Routledge that I might be the person to tackle the topic.

Thanks to Robert Upton and Cliff Hague of RTPI, and Helen Ibbotson of Routledge, for being so encouraging throughout the process of taking the book from concept to product. I found Cliff Hague's feedback on the manuscript most helpful.

Last, but never least, I owe an enormous debt to my family – Marty, Sari, and Caleb – for putting up with my obsession, forgiving my research absences, and for giving me priority at the computer when I needed it. They deserve medals for their patience, love, and understanding. Thanks also to Marty for editorial suggestions.

Any errors, omissions, or misinterpretations that survived the editorial process are my responsibility.

LIST OF TERMS AND MEASUREMENTS

APA	American Planning Association
brownfield	In North America this often refers to an industrial site available for redevelopment. In the UK, brownfield sites include abandoned industrial and former institutional sites
CBD	Central Business District, or downtown
CCTV	Closed circuit television, or video surveillance
CEU	Council for European Urbanism
charrette	a workshop held over several days in which interested parties work with designers to come up with design concepts for development
CIP	Canadian Institute of Planners
CMHC	Canada (Central) Mortgage and Housing Corporation (a crown agency of the Canadian government that conducts research and helps to finance housing and town planning)
CNU	Congress for the New Urbanism
DPZ	Duany Plater-Zyberk & Company (a Miami-based town planning and architectural firm)
form-based codes	a set of rules adopted by a local government to frame the design of a project. These may replace or augment zoning rules and design guidelines
gentrification	a process of neighbourhood change that sees more affluent households buying and renovating older districts, and thus displacing poorer households
grayfield	a commercial or institutional building or complex available for redevelopment
greenfield	an area previously in non-urban uses such as agriculture being converted to urban or suburban use
HOPE VI	Housing Opportunities for People Everywhere (a programme of the US Housing and Urban Development department that provided funds for the demolition and renewal of public housing in the 1990s)
HUD	Department of Housing and Urban Development of the US government

neo-traditional design	new design expressions rooted in traditional or classical patterns
ODPM	Office of the Deputy Prime Minister (the agency responsible for planning in England)
RIBA	Royal Institute of British Architects
TOD	transit-oriented design (a movement associated with Peter Calthorpe that proposes structured developments around transit stations)
UCA	Urbanization Control Area (Japan)
ULI	Urban Land Institute (an organization that conducts research and prepares reports that assist developers in the USA)
UPA	Urbanization Promotion Area (Japan)

GEOGRAPHIC CONTRACTIONS

American States: CA California, DC District of Columbia, FL Florida, MD Maryland, OR Oregon, TX Texas

Canadian Provinces: AB Alberta, BC British Columbia, NS Nova Scotia, ON Ontario

METRIC / IMPERIAL CONVERSIONS

1 hectare (ha) = 2.47 acres
1 square metre (sq m) = 9.29 square feet
1 metre (m) = 3.28 feet
1 kilometre (km) = 0.6214 miles

PART 1

THE RISE OF NEW URBAN APPROACHES

CHAPTER 1

NEW URBANISM(S) ASCENDANT

PARADIGM SHIFT?

In an address to the Royal Institute of British Architects (RIBA) in 1982, Leon Krier railed against modernism. If he were the head of RIBA, Krier vowed, he would plaster over the memorials the Institute had given to modernist architects whom he accused of destroying European cities and culture (Krier 1984a). Two decades later, RIBA elected a president who promoted new urban approaches, and the British government had committed itself whole heartedly to an 'urban renaissance'. How do we explain the transformation of dogma not only in the United Kingdom but in the United States, Canada, and several other countries as well? How did modernism find itself supplanted by a new urbanism that seeks inspiration in our urban past? Where did the traditional urban revival come from, what does it advocate, and why has it become so popular? These questions permeate this book.

The new urbanism involves new ways of thinking about urban form and development. Drawing on historic lessons from the most beautiful and successful cities, new urban approaches affirm the appeal of compact, mixed use, walkable, and relatively self-contained communities. Instead of car-oriented development practices, new urbanism argues for traditional architecture and building patterns that facilitate walking and that create strong urban identities. In sum, in an era when modernism has profoundly affected the shape of the city, new urbanism presents a new image of the good community.

This new way of planning and thinking about the city has travelled quickly, gaining the ears of practitioners and decision makers on several continents. In some ways, the dissemination of new urbanism parallels the rapid spread of garden city ideas a century earlier. New urbanism has become extraordinarily influential in the USA and Canada, making its way into suburban forms in many regions, and even supplanting conventional zoning in a few areas. Urban villages have made an impact in the UK and Europe, with government programmes to encourage innovation in many projects. We also find experiments in countries such as Japan, Australia, New Zealand, Turkey, South Africa, and Nicaragua. Although not every country uses the 'new urbanism' label for its activities, the Congress for the New Urbanism has given several international projects awards as good examples of the movement (e.g., CNU 2003, 2004b).

One of the key questions raised by new urbanism is its transferability. Drawing on a particular cultural history for its iconography, the movement looks to classical European cities, vibrant ethnic neighbourhoods of early twentieth-century America, and small towns for inspiration. Are its insights universal, or are they culturally situated in a particular time and place? The proponents of new urbanism have developed charters that present their principles as universal and timeless. New urban approaches have become so deeply intertwined with the strategies necessary to keep cities competitive in an era of globalization, that a high level of consensus supports their application internationally.[1] Yet some critics might challenge the suitability of employing these ideas so widely.

I am particularly interested in exploring new urbanism in theory and practice. In doing so, I use the term new urbanism in a broad sense, including American conceptions of 'The New Urbanism' but also encompassing related new urban approaches used in other countries that similarly draw on historic precedents and principles for inspiration. I want to investigate how new urbanism considers planning theory in its practice. Beauregard (2002:184) says that 'New Urbanism is both an urbanistic practice and a theory of urbanism'. How is new urbanism developing theory, and how does it contribute to our understanding of theory? How is planning theory responding to new urbanism and the insights its practice generates? Does new urbanist practice live up to its own theory? Are projects achieving the aims of the movement? Where do the challenges to creating good communities through new urbanism reside?

During the last two decades, new urban approaches have generated hundreds of books and thousands of articles. It has become impossible to keep up with the burgeoning literature in the field. Some well-known celebrities, including the Prince of Wales and Canadian billionaire Galen Weston, have taken a keen interest in promoting new urbanism. The mass media has profiled the movement in magazines, newspapers, and television. Some new urban communities are spectacularly beautiful and telegenic places (see Figure 1.1). Seaside, Florida, even made it to the big screen as a stage set for the movie, *The Truman Show*. Hundreds of consulting firms now specialize in new urbanism, and plans are being rewritten in many places to support new urbanist principles.

What drives this interest in new urbanism? Some express it in a word: sprawl. The twentieth-century city seems to have no limits, oozing inexorably over the landscape with little form or character. Suburban gridlock and a growing dissatisfaction with the placelessness of twentieth-century development put urban form back onto the public agenda. The former mayor of Toronto, John Sewell (1977), expressed the sentiments of many others in decrying 'jumblurbia': the monotony of uniform car-oriented suburbs, and of commercial strips that look the same everywhere. Globalization has generated what Kunstler (1993) calls the 'geography of nowhere' and Rowe (1991) sees as an undefined middle landscape: a 'sub'urban environment that is less

Figure 1.1 Handsome civic spaces
Attractive civic buildings, such as the community centre pavilion in Fallsgrove (Rockville, MD), epitomize new urban approaches. New urban communities are often extraordinarily photogenic.

than urban, yet far from rural (Duany et al. 2000). The mandate of new urban approaches is to facilitate the search for character and identity, and to repair the ailing landscape (Kelbaugh 2002).

The resentment of sprawl has spread quite far. Our popular culture seems to accept the premise that the suburbs are meaningless places. The suburbs – home to the majority of urban dwellers in many nations – have become the butt of jokes, and the locale for tragic movies. Worse still, the suburbs find themselves accused of generating social ills from anomie to road rage. Few stand ready to defend the suburbs.

As state agents charged with helping to generate better communities, planners by necessity search for appropriate urban strategies. In recent decades, planners have shouldered the blame for many of the problems of the city. In their efforts to find positive alternatives to suburban sprawl, many have found considerable appeal in new urbanism. Planners need theory and practices that offer a way out of a downward urban spiral, methods that respond enthusiastically to the challenges of growth. New urban approaches present hope for a better future.

The search for an alternative paradigm for urban development goes back to at least the 1970s. The energy crisis and accumulated government debt of the 1970s led to fiscal conservatism and retrenchment in the 1980s. David Harvey (1994) says that the end of the era of industrial accumulation spawned a period of flexible accumulation along with a new urban crisis. Certainly the fortunes of cities had begun to change, with many industrial cities showing clear signs of decline. Historic structures threatened by destruction became sites of contention and dispute: rallying points for a new approach to development.

The population was also changing. Many of the premises that had supported the ascendance of the garden city model in the early twentieth century no longer held.[2] Households were getting smaller, and often included two working adults. The population was aging rapidly. The cost of housing had become a significant barrier to many families. By the 2000 census in the USA, 25 per cent of households had a single person; half the population was 35 years or older. The same thing, or worse, was happening in other societies as well. Smaller, older households would need a different kind of city (Chiras and Wann 2003).

Into this opening – where the modernist city found itself challenged on several counts – stepped the vigorous prophets of new urbanism. As an antidote to the placeless suburbs, they offered a new prescription for neighbourhoods that followed historic principles and buildings that employed traditional materials. To reduce the ailments generated by car-oriented development, they advocated urban living in vibrant, connected, and diverse places. Their ideas have inspired a generation of designers and planners.

Despite the burgeoning consensus about new urban approaches among practitioners, we also find growing criticisms. The modernists' disdain for nostalgia is understandable: they see glorification of the past as naïve or wishful thinking. Often committed to finding innovative strategies for overcoming inequities, the modernists see no hope in efforts that failed previously.

Some might argue that new urbanism ignores complex urban realities. Indeed, new urbanism in many ways facilitates suburban development by making growth more attractive. Zimmerman (2001) argues that while it claims to promote urban lifestyles, new urbanism in fact legitimates growth on the urban fringe. As Marcuse (2000) says, new urbanist developments are not new and not urban. In some ways, new urbanism contributes to the problems of the suburbs: for example, its costs are high, making housing less affordable (Bookout 1992b). Making suburbs pretty does not undo injustice or stop sprawl.

New urbanism successfully found catchy concepts for packaging its message. For instance, its advocates have used the concept of 'Euclidean zoning' so extensively that one now hears the term regularly in planning discourse. Putting the name of a Greek mathematician in front of the word zoning sounds scientific.[3] Of course, in this case 'Euclid' refers to the US Supreme Court decision that approved the use of municipal zoning tools in Euclid, Ohio, in 1926. In using the adjective 'Euclidean', the new urbanists effectively dismiss zoning by making it appear overly rational and ancient. Similarly, the development of a smart growth agenda proved brilliant. By defining its own solutions as 'smart' the new urbanists have drawn the clear inference that other choices are 'dumb'.

A strong streak of environmental or spatial determinism runs through new urbanism (Harvey 1997). In this it follows many paradigms that have had immense influence in planning practice. Certainly the garden city revealed deterministic assumptions in its normative model. Its advocates believed that building satellite cities could control sprawl, protect agricultural land, safeguard the family, and eliminate the ills of the industrial city. Garden city ideas appealed widely and travelled to many countries to inspire planners during the twentieth century (Ward 1992). Of course, practice demonstrated that planning garden cities did not solve the problems of the city: indeed, we might argue that the garden city solution generated the problems that now inspire the new urbanists. Can we expect new urbanism to be similarly successful in being applied cross-culturally, and similarly unsuccessful in curing the ills of the city?

As new urbanism travels we see it transforming. Already we can argue that there is not a single new urbanism but rather many new urbanisms. Thompson-Fawcett (2003b) has documented clear evidence of a sharing of influences back and forth across the Atlantic and beyond. As with the earlier case with the garden city paradigm, the principal proponents of contemporary planning practice from Europe and North America engage in constant exchanges of ideas and debates. Leon Krier and Andres Duany appear to enjoy a rapport that has significantly affected their work.[4] In developing theory and planning principles, urbanists on both sides of the ocean build on each other's experience and insights. Practice may, however, take different forms. 'New Urbanist praxis has not become universalized over time and space; distinct variations exist' (Thompson-Fawcett 2003b:268). Hence at times I use the somewhat broader term 'new urban approaches' to signify that while the practitioners may not call what they do new urbanism, and while we see regional differences in emphasis, we do find many commonalities in the principles and visions that support contemporary planning practice in a wide range of contexts world-wide.

WHAT'S IN A NAME?

New urban approaches appear under a variety of names. The early projects done by Andres Duany and Elizabeth Plater-Zyberk were often called *neo-traditional town planning* or *traditional neighbourhood design* (TND). Peter Calthorpe and Doug Kelbaugh are known for *transit-oriented design* (TOD), *transit villages*, and *pedestrian pockets*. Since around 1993, with the development of the Congress for the New Urbanism, these approaches have fused in the *'New Urbanism'* (usually written with capital letters by its proponents). By the mid to late 1990s, many people were talking about *urban villages* as nodes of new urban development. The National Governors' Association in the USA used the term *new community design*, and Emily Talen (2001) tried out *traditional urbanism*, but those did not catch on in a big way. By the late 1990s, the British grew excited about an *urban renaissance* and launched urban village programmes, while the Americans and Canadians signed on to *smart growth*. In an era where branding has become the key to marketing success, the new urbanists have been successful in establishing their solutions as the strategies for achieving better places.

Whatever the label used, these new urban approaches share common principles: fine-grained mixed use, mixed housing types, compact form, an attractive public realm, pedestrian-friendly streetscapes, defined centres and edges, and varying transportation options. In many cases – although not universally – they favour traditional architectural and design patterns, open space networks, and connected street layouts.

New urban projects are appearing with increasing frequency in many countries. Each year, the *New Urban News* (produced by the CNU) reports with pride on projects proposed or underway. Thousands of people now live in new urban communities: many people find them beautiful and meaningful living environments. Are the places that have been built examples of good communities? Not everyone believes so. Scully (1991) implied that they might be a new suburbanism. Leung (1995) thought they were a new kind of sprawl. Barber said 'sprawl with trim' (1997:A2). Pyatok (2000:814) suggested 'a more seductive form of business as usual'. Baxandall and Ewen (2000:251) criticized a form that they saw as the 'fantasy theme park village', while DeWolf (2002) lamented what he thought were *faux* towns. Shibley (2002) damned them as Potemkin villages. Marshall (2004) called them suburbs in disguise. In earlier work I also challenged the ability of new urbanist developments to make a significant difference. I argued that, like the garden city paradigm before it, new urbanism was still producing 'cookie-cutter' suburbs: stamped out in uniform patterns (Grant 2002a). Getting projects built does not necessarily mean achieving the principles that new urbanism promotes.

Indeed, where we could argue that the war on blight produced suburbs on steroids, we might say that the war on sprawl is producing suburbs in period costume. The bulked up suburbs of the modernist era led to significant long-term health and economic implications; the dressed up suburbs of the contemporary period mask continuing inequities and unsustainable behaviours.

With his characteristic acerbic wit, Duany dismisses such criticisms. He and other new urbanists see academics as threatened by new urbanism, and find the attacks unconvincing (Ellis 2002). Undoubtedly there is more than a grain of truth in that observation. With success comes criticism, some based on envy of a paradigm that has enjoyed such ready popularity. Some criticism reflects discomfort with the premises and assertions, raising theoretical objections to the model. As examples of new urbanist developments spread, an increasing measure of criticism comes from studies of practice that ask whether new urbanism can actually achieve its aims. In response to the critics, new urbanists have begun a campaign to stake territory in theory and to document the successes of the projects built.

REPRESENTATIONS OF NEW URBANISM

In some ways we can see the history of planning as a series of new urban approaches. Crises in urban conditions lead to new planning concepts and approaches meant to rectify the situation. How do we situate the origin of the new urbanism of the late twentieth century?

Was the first salvo of new urbanism from DPZ, Duany and Plater-Zyberk's planning and design firm, in 1982, with the creation of the Florida resort at Seaside?[5] Or should we look earlier to Krier's (1978) writing and lectures about urban quarters? Perhaps we might date it to the end of modernism with destruction of the Pruitt-Igoe public housing high-rises in St Louis in 1972. Or we might look even earlier at the influence of Jane Jacobs's (1961) short but provocative best-seller on the fate of American cities.

We could look much further back in history for inspiration. After all, Kelbaugh (1989) notes that new urbanists draw on 2000 years of experience in building good cities. The first planned cities in the Indus Valley date to as early as 2500 BC (Hammond 1972). They feature some of the principles that new urbanists employ today: mixed use, small grid blocks, pedestrian orientation. The lessons of ancient cities give us a sense of the range of approaches to urbanism through history: varied ideas of what the city can be or should be.

New urbanism takes a selective look at history, drawing its lessons primarily from the classical traditions of the Greeks, Romans, and Europeans. Urbanism in these traditions typically facilitated individual and cultural aggrandizement. The Greeks and Romans, and later colonizing European nations, built planned settlements as a way of achieving individual or imperial ambitions. Settlements controlled space. In these examples, classical principles served the interests of power. Urban form became a vehicle for conditioning submission.

In finding inspiration in the 'timeless ways' of classical forms, new urbanists rarely consider the cultural and social context in which their treasured principles developed. They abstract the architecture from its setting and social meanings. They value the aesthetics of classical forms while they focus on trying to meet the needs of contemporary urban residents.

REPRESENTATIONS OF HISTORY

History constitutes an integral component of new urbanist theory and practice. New urbanism uses history for inspiration. As every historian knows, of course, history involves selection and interpretation. Contemporary experience and our aspirations for the future inevitably frame our understanding of history. We do not write history on a blank slate, but on a parchment created from what we know and love.

In writing history, then, we link our past with our present and our future. Our interpretation of history can become a justification for a break with the present to influence the possibilities of the future. Similarly, arguments we might have about the future often involve projections of the past.

The history we know, the stuff of text book and encyclopaedia, is typically a history of tyranny. Historians have documented and celebrated the dominance and power of great men and empires that succeeded on the blood, sweat, and tears of others. The classical empires insisted on order, discipline, and hierarchy both in social and spatial arenas. They used colonization, militarization, and commodification to achieve their aims. They built great cities to glorify their achievements. These cities became the epitome of civilization as we know it.

When Christopher Alexander (1979) talks about timeless ways of building, he looks for historical and cross-cultural examples that reflect beauty and functionality. His work suggests that past, present, and future blend seamlessly together when it comes to good design. New urbanism shares this faith in universal principles. Yet a study of the incredible diversity of urban forms through history dispels any illusion that all civilizations share an understanding of the quality or meaning of urban forms. The architectural (archaeological) record reflects diverse regional styles linked to materials, functions, social order, economy, climate, religion, and values. Efforts to generate timeless principles are necessarily selective: they reflect the values of the time and context of those who extract them. They are thus a product of ideology and practice.

REPRESENTATIONS OF PRACTICE

New urbanism developed as a movement within practice as a response to the failure of cities and suburbs in the mid-twentieth century. In the wake of Jacobs's (1961) criticisms of planning, planners increasingly felt a level of angst over, or even responsibility for, the condition of the city. Schooled in preparing and implementing plans and land use regulations, they appreciated the appeal of a new movement that offered a simple, tangible, and marketable recipe for practice. New urbanism described clear prescriptions for action with a promise of improved outcomes.

Lynch (1981) explained the ways in which physical conditions contribute to satisfaction. Since the early twentieth century, planning practice has placed a high priority on enhancing urban conditions both for the economy and for urban residents. Planning practice is intricately caught up in place. Planners work in a context where place matters (Graham and Healey 1999). In the post-industrial economy, globalization requires communities to improve the quality of local places to compete for capital and labour. We have shifted from a modernist paradigm in which we concentrated on function over form to a practice that increasingly privileges form as a mechanism to attract investment. The concern about resident satisfaction that animated Lynch has yielded to a calculus more likely to favour the entertainment quotient of a place.

Foucault (1977:141) wrote that: 'In the first instance, discipline proceeds from the distribution of individuals in space.' Today a small international élite has considerable control over the economy, public tastes, and location choices. Their decisions affect the fate of cities. Within our cities we often find residents' life chances differentiated by neighbourhood. The most affluent areas are carefully controlled and beautifully appointed. These include the gentrified districts, the gated enclaves, the upscale suburbs. Social, economic, and physical barriers keep poorer members of the community from feeling welcomed in these areas. 'That which disguises itself as a disinterested, friendly, hospitable consumption sphere in practice draws up dividing lines between those in control and those they are excluding' (Douglas and Isherwood 1996:109). New urbanist principles not only generate attractive homes for affluent consumers, they increasingly mark landscapes of inclusion and exclusion.

Planning operates within a social structure that defines the public interest as a justification for state intervention. That is, as long as restrictions on the use of property may be shown to have a wider public purpose, then planning is a legitimate practice. The modernists accepted that planning to ensure greater efficiency in the use of resources was in the public interest. Many early modernists also had socialist leanings and promoted equity. Practitioners felt quite comfortable in asserting the public interest in their work.

Urban renewal began to undermine faith in a unified public interest during the 1960s. Civil rights advocates and feminists articulated alternative interests that planners could no longer ignore. Hence we saw challenges to the public interest criterion in planning theory after this period (Klosterman 1980). Contemporary planning theory acknowledges and seeks ways to accommodate diverse voices and needs in planning practice (Sandercock 1998).

By contrast, the new urbanists have revived a commitment to a notion of the common good. For them, beauty, order, and coherence are universal principles of urban quality. New urbanism may appeal to practitioners because its rules for practice allow planners to operate without worrying about the messiness of divergent claims. Narrowing the focus of inquiry to a range of issues where correct answers are presumed to exist simplifies practice. Thus the public interest is redefined in new urbanism as coterminous with a definition of the good community as a particular kind of place. Practitioners searching for strategic solutions to the problems their communities face have thus found useful and persuasive answers in the principles of new urbanism.

REPRESENTATIONS OF CULTURE

Just as history gives us greater depth in understanding the problems of cities and possible strategies for dealing with them, so can investigating practice in other cultures offer useful contrasts. While focusing on a single regional or national practice may reveal common patterns, cross-cultural studies illuminate difference. Understanding the experiences of other cultures presents alternative explanations of what can happen and why. Looking at new urban approaches employed in several countries allows us to explore the role of values, history, and practices in these diverse contexts.

The East and West have quite different ideas of urbanism (Shelton 1999). With their basic functionality of providing shelter, even homes reflect and reinforce diverse values associated with domesticity, individuality, and privacy. Thus, for example, we often find new American homes with an open plan for living areas around large kitchens for entertaining. The English home closes off rooms so all functions may be separated, keeping guests to public parlours. The typical new house in Japan combines a living room, dining room, and kitchen in a multi-purpose family room that rarely hosts visitors. We might argue that the American home facilitates display, the British home promotes privacy, and the Japanese home accommodates family intimacy (Ozaki 2002). Similarly divergent values and views operate at the larger urban scale: streets, courtyards, and parks can mean different things in disparate cultures.

Our society simultaneously embraces diversity and resists difference. The institutions of our public culture – as Horne (1994) calls our schools, museums, and news outlets – celebrate diversity. In Canada, for instance, the government institutionalizes multi-culturalism with events and activities to acknowledge its rich immigrant heritage. At the same time, however, the popular culture – the activities, products, and ideas generated outside of state channels – promotes homogeneity. Advertisers encourage consumers to adopt common products as status markers; entertainment programmes disseminate messages about fitting in that influence young viewers; residents' associations fight unwanted uses and peoples in their communities; gangs enforce group colours and expectations among members.

Day (2003) argues that new urbanism has underestimated the significance of diversity and has developed few strategies to accommodate difference. New urbanism seeks to identify universal principles. Universal principles assume commonality. Efforts to generate a widely-applicable lexicon of good development reinforce accusations that proponents overlook difference of needs and means. The challenge for new urbanism is to develop a wider body of lessons from practice to help identify whether universal principles may apply.

Cities are spaces of cultural consumption in which goods mark status (Douglas and Isherwood 1996). In the contemporary city, history and community have become commodities. New homes are sold as place products through which consumers identify their values and their status. Harvey (1989:271) suggests that urban life in the post-modern city is an 'accumulation of spectacles'. Cities act as display and performance spaces. Marketing of urban and suburban neighbourhoods depends on product differentiation. In this view of the city, new urbanism adds value to particular neighbourhood products. It allows consumers to mark their status and identify their values for all to see.

As new urban approaches travel into different contexts, they inevitably adapt to regional and local cultures. We will consider how some ideas associated with new urbanism may have been picked up and transformed in translation. In Part 2, I discuss new urbanism in its original American context, and then look at the application of related ideas in Europe, East Asia (with a focus on Japan), and Canada. While new urbanist ideas translated readily into the Canadian context, the transference proves more partial in the UK and Europe, and quite tenuous in Japan.

Calls to change the way our cities work may not recognize that existing forms are deeply entrenched. As Filion (1996) notes, reorienting growth away from auto-oriented patterns is incredibly difficult. Support for the compact city model is almost universal, but implementation remains a challenge. If we empower residents to decide, then they tend to stick with 'Fordist consumption patterns' (Filion 1996:1653). Hence, the new urbanists argue, we must give designers more authority to identify and implement appropriate responses.

When new urbanist discourse discusses culture, it reveals its proponents' historic and class biases. Krier (1978:40) writes that 'popular culture and intelligence have been destroyed almost beyond the point of redemption'. He suggests that high culture – capital 'C' Culture – was damaged by the bourgeois search for power in the twentieth century. In this view, design represents an embodiment of Culture, with an important mission, as Van der Ryn and Calthorpe (1986) say. The clear implication is that the masses lack the ability to recognize quality but must rely on the expertise of the designer to enlighten them. While the anthropologist sees culture as carried in the ideas, practices, and artifacts of ordinary people, the new urbanist defines Culture as the ideas, practices, and artifacts of the élite (often of an earlier era). This interpretation confers considerable power on the designer, with significant implications for planning theory.

REPRESENTATIONS OF THEORY

The planning profession needs theory to inform practice (Friedmann 1979). Theory provides justifications for decisions, offers guidance on possibilities, and facilitates ethical behaviour. Theory attempts to extract rules to describe, explain, and predict the world. We draw on our interpretations of the past, our present, and potential futures to develop theory. By and large, though, we can see theory as a product of the times, embedded within a particular social context (Grant 1990). Moreover, despite our wish to build comprehensive theory, the theory we use as planners is often partial. Planning occurs in such complex conditions that we cannot possibly explain everything. Of necessity, we selectively consider factors and construct premises. The issues on which we focus, and the choices we make in representing reality, reveal the values we hold as theorists.

Theories in planning respond in part to the challenges of practice. For example, the garden city model presented an approach to try to control growth by harnessing it in manageable chunks in satellite new towns. Garden city advocates hoped to reconcile the city with nature, and to improve the conditions for working class households. New urbanism responds to issues of sprawl and modernity in the city. It seeks to manage growth by making cities more urban/urbane, and to restore the kind of vibrant neighbourhoods experienced in the early twentieth century.

We find the roots of new urbanist theory most evident in the work of Jane Jacobs and Leon Krier. Jacobs (1961) focused on ensuring a mix of uses and people in the city. She documented the failures of modernist planning ideas – of high-rise buildings and large parks – to maintain the vibrant, fine-grained mixed use of the ethnic neighbourhoods of Greenwich Village that she loved. She associated deteriorating civility with changes to urban form. Krier (1978) concentrated more on design questions. His view of the good city insisted on visual coherence but mixed functions. He looked to compact pre-industrial cities as models of places that integrate urban functions while avoiding the confusion generated by a mix of building styles or the inclusion of non-urban elements (Krier 1984a).

Kevin Lynch devoted a considerable part of his career to trying to offer guidance on good urban form (e.g., Lynch 1981). He described three kinds of theories that explain spatial patterns in the city. Planning theory or decision theory examines how decisions about city form are made, and how cities take shape. Functional theory explores the ways in which form works. Normative theory connects form to values: it answers the question of what makes the city good. Contemporary theory in new urbanism is essentially interested in normative issues (rather than questions of process or function). As Talen and Ellis (2002) say, planners need guidance on what is good in the city so that they can make better choices. What is the shape and character of the good city? New urbanist discourse has invested heavily in answering that question.

By contrast with the questions that captivate new urbanists, concerns about processes and functions loom large in what is more typically thought of as 'planning theory'. Mainstream theory explores issues of spatial patterns and causality (Cooke 1983). Its theorists ask: how does the city come to take the shape it does? How do the shape of the city and the processes that generate it affect the life circumstances of citizens? Of course, planning theorists are also interested in what makes a good city, but they are more likely to define 'goodness' in terms of equity, democracy, or social justice, rather than beauty, harmony, or balance.

Practitioners who read conventional planning theory for guidance might reasonably complain of élitism in the academy. In recent decades, the better known planning theorists like John Forester have tended to turn to remarkable European thinkers like Jurgen Habermas or Michel Foucault for inspiration. While all these men offer brilliant insights, they write in a code that is often indecipherable to the average reader. University professors initiate graduate students in the jargon, but the lessons seldom influence practice. As Fischler (2004:363) notes 'jargon-filled prose is more likely to estrange' than attract practitioners. Hence Judith Innes (1983) could write of a widening gap between theory and practice.

In her 2000 paper on planning theory, Fainstein outlines three common approaches in contemporary planning theory discourse: the communicative or collaborative model; new urbanism and neo-traditional approaches; and the political economy model. The first focuses on process, the second on design outcomes, and the third on equity outcomes. We see in these theories quite different models of the good community: one where justice prevails, one where beauty prevails, and one where equality prevails.

While the themes that animate planning theory are of critical importance to planning practice, they do not always engage practitioners. What key themes pervade planning theory? Since critics exposed the weaknesses of rationalism in the 1960s, theorists have been anxious to find alternative models to inform practice. Current attempts to go beyond the rationality of modernism to find ways to accommodate diversity and democracy have generated interest in issues of communication, collaboration, and story telling. Fainstein (2000) describes two major approaches in this literature. American pragmatism, following the early twentieth-century sociologist John Dewey, focuses on studying behaviour and finding the best practices to emulate. Communicative rationality borrows from Habermas and attempts intersubjectivity (trying to get into the perspective of the other players). These postmodern approaches converge in their efforts to provide guides to action. As Forester (1989) explains, planners must be honest, open, and inclusive as they work with others in searching for consensus. In that sense, then, communicative planning theory has a normative dimension to its prescriptions for process.

Another key issue in planning theory is that of power. Cities are spaces of power and control. Whose interests are furthered by planning and design activities? Who decides what is good in the built environment? What are the power politics of a design movement that claims it can define the characteristics of the good community? What are the spatial ethics of planning (Upton 2002)? As Harvey (1989:265) says: 'Different classes construct their sense of territory and community in radically different ways'. In general, though, the affluent prescribe and command space. A phenomenon like gentrification entails class change in neighbourhoods as the value of land improves and residents with lower incomes are forced to find housing elsewhere. Conditions of flexible accumulation prevail in the contemporary city, with rapid changes in the use and value of land. Political economy analyses of the city see space as reflecting the 'ideology of the ruling groups and institutions in society' (Harvey 1973:310). In this view, the good city would instead allow the labouring classes to reap an equitable share of the wealth of the city.

Fainstein (2000) argues that all the contemporary theoretical models react against the failures of rational planning. Most of us understand that objectivity is impossible. We accept that planners cannot be comprehensive or omniscient. Many agree with Harvey (1989:3) that 'Science can never be neutral in human affairs'. Yet giving up on the rational planning process as an ideal is a challenge for many practitioners. Not everyone is persuaded that if objectivity is impossible, then explicit subjectivity is the only other option. At the same time, though, much of the theory currently available, both in the mainstream of the academy and in the new urbanism literature, proves explicitly normative: it offers prescriptions for action.

Issues of power and communication hold little interest for new urbanists, even as their practice engages these concerns. While they may not deny the contributions of mainstream planning theory, new urbanists search for theory for different reasons. They want principles and premises that can inform or justify the choices they make about the shape of the city. Their questions are not of process, but of materials and dimension: what kind of city shall we build? In the absence of guidance from the gurus of planning theory, the new urbanists have turned to developing their own theory. In their search for suitable models, they have found utility in ecological theory – albeit a narrow and, some may say, distorted version of the discipline.

If objectivity is impossible, and increasingly practitioners are expected to take a moral stance, then planners need reasonable theory to substantiate their normative positions. What does a normative planning theory that can work in practice require? Planners generally accept that good practice relies on open and honest communication and democratic processes. Lynch (1981) says that a theory of good urban form should be simple, flexible, and easily applied. Planning the good community needs a theory of possibilities rather than iron rules. It must deal with plural and conflicting interests. It proves a tall order to find theory that can meet these criteria and generate consensus within the profession.

CONNECTING THEORY AND PRACTICE

Contemporary planning theory takes a keen interest in practice, although authors sometimes complicate the discussions by speaking of *praxis*.[6] Whatever label they use, though, theorists see practice as a source of important lessons that can inform theory. Practitioners may not reciprocate with such passion for viewing theory as a tool for practice. Although planners believe that the profession needs good theory, they often express frustration about the difficulty of using the theory they learned in planning school. While I seldom hear practising planners talking about Habermas, I do find them discussing Andres Duany or Jane Jacobs. This may reflect a level of comfort and sympathy with the earthy language and heart-felt opinions of some of the proponents of new urban approaches; it contrasts notably with the squeamishness that practitioners can express in trying to wade through thick theory. Similarly, we are less likely to see the media interview John Forester, Patsy Healey, or David Harvey about their latest interesting article than we are to come across interviews with Peter Calthorpe or Leon Krier in front of their latest development projects. Communication theory is not telegenic; political economy theory is leftist; that leaves new urbanist dogma to grab the spotlight.

In its early days, new urbanism seemed disinterested in theory (Shibley 2002). Calthorpe explicitly privileged practice over theory: 'Because the social linkages are complex, the practical must come first' (1993:10). At times in Calthorpe's work we find a kind of pride in avoiding the theoretical. 'The realities of the modern American city require a model which incorporates and reconfigures the diverse uses at work in the marketplace, not a theoretical construct which hypothesizes a fundamental change in the architectural "building blocks" of development' (Calthorpe 1993:45). Bressi (1994) argued that theory could cloud issues for an action-oriented discipline.

Amanda Rees (2003) describes new urbanism as anti-intellectual. Bohl (2000:777) agrees: 'the New Urbanist literature has not involved social scientific theory building and empirical testing, but rather marketing and manifesto instead'. Many key new urbanist authors write without citing sources, or building on previous planning literature: they engage in a kind of armchair philosophizing that has disappeared from most disciplines. Discussion lists amongst new urbanists are replete with rhetoric and even name-calling, sometimes displacing efforts to develop well-considered analysis or theory.

Despite the disclaimers and scepticism about theory, new urban approaches *are* informed by theory: sometimes that theory is explicit, but more often it remains implicit. New urbanism builds on selective precedents, and reflects a particular way of thinking about the city. In recent years, its advocates – especially those in the academy – have tried to establish a theoretical base for new urbanism. They recognize that new urbanism needs theory in order to gain respect in the academy and to increase its influence with decision makers. Theory can help new urbanism offer guidance grounded in a broader logic as it identifies the principles of the good community. The theory that informs new urbanism proves explicitly normative.

Although we can see new urbanism in part as a reaction against the extreme rationalism of the modernist approach, in some ways new urbanism continues the modernist project. The rational approach thought progress was possible only through planning (Boyer 1983). We find echoes of that in new urbanist writing. For instance, Kelbaugh (1997:112) says that 'cities that are not planned in some manner end up as illegible and confused as Houston or Tokyo'.[7] Planning creates order and coherence in the built environment. Good designers and planners have the expertise to restore good communities.

Of course, the new urbanists also argue that over-planned or badly planned cities end up sterile or incoherent. In these cases, zoning is the usual problem, separating uses and forcing people into cars for mobility. Instead of zoning, the new urbanists prefer codes and covenants that set design rules to ensure harmony of built form. In some ways, the new urbanist search for rules to ensure order aims to be even more all-embracing than zoning could ever be. Design codes may have rules for everything, down to the type of waste receptacles and the size of signs. The coder sets rules for posterity.

Like rationalism before it, new urbanism affirms the importance of expertise. New urbanism posits two types of planners. The bureaucrat/administrator writes policy and enforces zoning rules: this planner has no imagination and little concern for urban quality. By contrast, the creative designer/planner generates visions of better futures and helps people achieve them. Cities need qualified designers to lead the design and planning processes, to identify and implement good forms and practices. Without expert designers, the new urbanists believe, the good community will remain elusive.

At first new urbanism was essentially an architectural movement to reclaim town planning. The major practitioners came from architectural backgrounds; few trained as planners. As the movement spread, however, more and more planners were 'drawn to the neo-traditionalist idea both out of a sense of guilt over past practice and a belief that planning could still play an important role in addressing social ills' (D. Hall 1998:28–9). In the process, we see that policy writing takes a reduced role in new urban planning practice. Short vision statements sometimes replace municipal planning documents that once ran to volumes. Local governments in some areas have hired urban designers and streamlined or transformed their zoning codes and land use regulations. Town planning has largely internalized the architectural critique, and absorbed many lessons from new urbanism.

Planners help communities explore options and reach decisions. If planning history and theory teach us anything, they demonstrate the need for caution in making assumptions that particular changes will solve urban problems. The garden city model did not end sprawl. Urban renewal did not eliminate poverty. The compact city will not solve every social ill (Breheny 1992; P. Gordon and Richardson 1997; Graham and Healey 1999). We must be careful to ensure that consensus about appropriate ends and means does not mean that élite groups dominate decision making: 'design-based solutions tend to narrowly crystallize the time-space requirements of dominant interests within the built form' (Graham and Healey 1999:641). If, on the one hand, the planner administers a democratic process to accommodate debate over ends and means, then the planner might play a socially-progressive role. If, on the other hand, the planner applies a normative framework that implements a particular vision of the built environment, then is the role a progressive one? The answer depends on whether one believes the vision itself is progressive or not. Therein lies the debate about new urbanism.

The new urbanists see the public interest and a conception of the progressive community as strongly linked to good urban form (Talen 2000a; Talen and Ellis 2002). For the new urbanists, planners must go beyond managing processes to take positions of leadership in communities. Equipped with a theory of good urban form, and the power of their convictions, planners would have the expertise needed to generate good communities. In many ways, the new urbanist position thus seems closer to modernist view of expertise than the designers might like to admit.

THEORIZING IN NEW URBANISM

New urbanism concentrates its development of theory around a few themes, with special attention to issues of community, organic analogies, and built form. Community is an important concept for new urbanism because in many ways it constitutes the ultimate goal of design interventions. New urbanism seeks to create opportunities for positive social interactions in space. It represents an effort to create local spaces for socializing: places to shop, educate, play, and work near home. New urban approaches typically envision bustling streets, with people hopping on streetcars, calling 'hello' to the greengrocer on their way home. Nostalgic views of intensely interactive small communities of times past animate the vision.

Like many planning movements, new urbanists subscribe to organic analogies with some frequency (Thompson-Fawcett 1998). In recent writings, new urbanists link their ideas to environmental and ecological theory. Describing the transect – the idea of a gradient of building types from most rural to most urban – the new urbanists use ecological terminology and sustainability concepts to try to develop a regional theoretical framework (Duany and Talen 2002a, 2002b). The principal concept that underlies the transect idea involves separating nature and city as if they occupied different ecosystem niches. Where the garden city sought to integrate town and country to ensure the best of both in a controlled way, new urbanism spurns such mixing of conceptual categories. For garden city planners, the scale and chaos of the city, and the loss of connection to nature, threatened the urban resident. In smaller satellite communities, nature would penetrate the managed urban fabric to restore health and happiness. By contrast, the new urban ideal separates nature and city. Each threatens the other. As Krier (1984b:30) says, 'city and countryside are antithetical notions'. The city must be distinctly urban, with natural elements formalized and specified. Boundaries that contain growth and define edges protect the countryside. Inappropriate natural elements in the city limit densities, and confuse the urban order. Rees (2003:102) argues, however, that 'the collapsing of social, cultural, and economic theory with ecological theory is fraught with danger'.

The new urbanists arguably have a limited understanding of ecological theory, and use it in selective ways. Contemporary ecological theory would challenge the notion that natural and cultural processes can be separated in the landscape. For instance, landscape ecology sees nature and culture as false separations. But the new urbanists seem wedded to dichotomies, anxious to keep distinct forms and functions in their place in the urban environment. The cautious adaptation of organic analogies serves their interests in articulating a framework for the good community.

The principles of form that suffuse the normative vision of the good city for new urbanists typically derive from classical elements and traditional (pre-automobile) urban patterns. Principles for good urban form deal with issues such as massing, materials, proportions, formal spatial relationships, and the creation of voids or solids. Harmony, balance, coherence, and imagability are the goals of good planning and design.

VALUING NEW URBANISM

New urbanism appeals to planners because it encompasses the range of values important to the profession. From the earliest days of town planning, reformers looked to municipal intervention in property development as a way to achieve four key values: equity, amenity, health, and efficiency. Theories that bridge and link these values stand a good chance of becoming popular. Of course, achieving these values can never be perfect. Practice reveals the inefficiency of some theories and movements to implement the values, especially equity.

Table 1.1 illustrates the key values operating in planning theory and practice, and shows their relationship to some of the movements of the last century. Utopian approaches are arrayed to the left of the table, with more technical approaches to the right. Some movements and theories have tended to emphasize one of these values much more than other values. Theories developed in the early years of planning often saw one of the values as primary. For instance, the Public Health movement advocated sanitary and housing improvements with health as its priority. Some more recent theories also promote one value over others: radical planning theory favours

Table 1.1 The key values of planning in planning movements

	Mainstream values		
Equity	*Amenity*	*Health*	*Efficiency*
socialism settlement house communes advocacy radical Marxism communicative action collaborative planning	city beautiful beautification parks movement traditional neighbourhood design	public health regulatory reform	urban reform movement public service movement scientific management resource conservation
	rational, comprehensive, master planning, regional planning, strategic planning, integrated resource planning, transit-oriented development		
garden city, neighbourhood unit, healthy communities, sustainable development, new urbanism, urban villages, smart growth			
UTOPIAN APPROACHES		TECHNICAL APPROACHES	

equity over other values; traditional neighbourhood design focused on amenity ahead of other themes. The table suggests that most mainstream movements of the mid to late twentieth century united the values of efficiency, health, and amenity.

Theories that integrate notions of equity with amenity, health, and efficiency are the most powerful ideological paradigms in planning. Their breadth as all-embracing approaches to achieving the good community makes them attractive to a wide range of practitioners and theorists. As practice soon demonstrates, however, theory that incorporates the widest range of values cannot necessarily deliver on its promise. Equity remains elusive, and even amenity proved in short supply as the garden city and neighbourhood unit became entrenched planning ideas transformed and simplified through practice. Will contemporary theories and movements with the same breadth fare any better? Can any theory or practice deliver equity, or is it an illusory goal?

SEARCHING FOR CIVILITY

In her classic book, *Purity and Danger*, anthropologist Mary Douglas (1966:3) wrote: 'The whole universe is harnessed to men's attempts to force one another into good citizenship'. The justification for modern town planning is in part the quest for a civilized and civilizing community. Having given up on ritual incantations and religious dogma to enforce good behaviour, we turn now to government apparatus and private covenants to set and enforce codes of conduct.

> The New Urbanism in fact connects to a facile contemporary attempt to transform large and teeming cities, so seemingly out of control, into an inter-linked series of 'urban villages' where, it is believed, everyone can relate in a civil and urbane fashion to everyone else.
>
> (Harvey 1997:2)

As we seek to restore civility, we find an appeal in the promises of new urbanism. We need to believe that we can minimize crime and maximize happiness. We long for a good city that is civilized, beautiful, socially-engaged, and just. We want to live in places that are proud of the past and hopeful about the future.

But how do we create a civil and just society? New urbanism can deliver on the desire for beauty, but can it restore civility and justice? Rabinow (1984) says that while Noam Chomsky suggests a just society will be guided by universal reason and justice, Foucault believes that justice must involve a struggle to change power relations. For the most part, planning theorists have tended to agree with Foucault: hence radical planning advocates overturning power régimes that harm the interests of the poor and push them into alternative behaviours. John Friedmann (1979) wrote that a just society will resist hegemonic power. Planning for justice would mean a radical form of practice as the moral option. Justice entails a programme of reform to improve living conditions for all, to enhance social mobility, and to guarantee democratic participation. Thus we see that ideas of the good community as a just society reflect the premises of the political economy approach.

While new urbanists want civility, they seldom show an interest in changing power relations as a strategy of achieving a good community. Instead they focus on issues of community character, identity, and sense of place. They believe in universal reason. They want to make everyday life more pleasant and comfortable. They seem content to take the political and social structure as a given, but within that they hope to create better places for people to live their lives. The new urbanists problematize the character of space rather than the social structure that generates it. They believe that an attractive and meaningful built environment can create conditions to enhance civility amongst citizens.

New urbanism also relies on surveillance to limit bad behaviour. One role of the state is to discipline and control individuals (Foucault 1977). Constant surveillance has become a feature of modern life. In a 2001 speech, British Prime Minister Tony Blair announced funding for more police and more video surveillance (CCTV) as part of an initiative to improve urban livability and keep cities competitive. While new urban projects may not rely on CCTV for social control, new urbanism advocates 'eyes on the street'. Designers are encouraged to use built form to discipline behaviour. 'The exercise of discipline presupposes a mechanism that coerces by means of observation' (Foucault 1977:171). Users of spaces are conditioned to good behaviour because they believe that others can see them. The good community keeps people in line.

If we had a civil society, would we need planning? In *The Drama of Democracy*, I argued that planning provides tools whereby we manage insolvable conflicts over the use of land. 'Our society has institutionalized community planning as part of a cultural apparatus for dealing with conflict and social control' (Grant 1994:5). As we expand the role of physical design planning, we increasingly use planning to generate the rules for the good city. We still lack consensus, but new urban approaches offer a normative framework that eases the process by promising more explicit definitions of appropriate ends and more effective mechanisms of social control.

SEARCHING FOR THE GOOD COMMUNITY

Lots of people have written about cities and what makes them good. Images of the good community sell books and magazines, and convince people to tune in to television programmes. We can identify a lengthy list of award-winning journalists and urban critics to prove the point: Lewis Mumford (1961), Jane Jacobs (1961), James Howard Kunstler (1993), Jane Holtz Kay (1997), and Alex Marshall (2000). Even royalty finds cities amusing: Prince Charles wrote a best-seller on his vision for Britain (Prince of Wales 1989). Everyone lives in settlements, so everyone is a potential critic or visionary.

Some scholars have given the matter considerable thought. Haworth (1963:22) says that: 'A good community is one which, by providing its members with a wide variety of opportunities for significant activity, encourages their growth, the development of whatever potentialities they possess.' For the philosopher, the good city allows self-actualization. Lynch (1981:116–17) seems to agree: 'that settlement is good which enhances the continuity of a culture and the survival of its people, increases a sense of connection in time and space, and permits or spurs individual growth'. Perhaps this may be a particularly Western view of the aim of a good community. Not all cultures value self-actualization equally, but most recognize that the physical elements of the city play a role in enhancing or undermining people's lives.

Before proceeding to develop a theory of good urban form, Lynch (1981:1) begins by suggesting that it may be meaningless to ask what makes a good city. Cities are simply facts of nature, he says. They are certainly facts of history. They are here; we have to deal with them. Settlements may be good for some and not for others. But if we are to justify our activities in planning communities – to account for the choices we make on what to do or not do as interveners in development decisions – surely we must have some notions of goodness in mind, either for processes or for outcomes?

As planners we focus on places, on the locations where social communities form in space. The spatial element is a given of our concern, and the social is implied. But how do we decide what is good? Often the models we look to as examples of good cities reflect a time when a small proportion of the population lived in cities. Now that more than 50 per cent of us live in cities, we think of our cities as flawed. Have we exceeded the carrying capacity of what we might define as the good city? Those beautiful and charming urban gems of history achieved their compact glory by extracting the wealth of their hinterlands. Peasants struggled in the countryside to keep the burghers of Europe in silks (Huppert 1986). The lofty spires and fine piazzas reflect an unjust distribution of costs and benefits that we have no interest in reconstructing.

We seem keen to try to reconstruct a city that is centred, transit-dependent, and mixed in use. That kind of city thrived at the turn of the twentieth century. As Rae (2003) explains, though, that form uniquely suited a particular time and place. A constellation of factors related to energy generation, transportation technologies, agricultural innovations, and population dynamics created the centred city. When technologies and circumstances changed, the city transmogrified. It may no longer be possible to restore the centred option.

New urbanism proves vulnerable to idealized notions of good urban spaces. 'Concepts like village and community are heavily laden with moral and emotive connotations of an older, natural social order' (Brindley 2003:58). Traditionally American and British cultures have idealized the rural; elements of that longing survive in contemporary ideas of urban villages. Yet new urbanism also reveals the lure of certain kinds of urban places. In planning a city in Iran, Robertson (1984:11) says he was 'trying to recapture the good city, the low-rise, harmonious, age-less City of Man'. This is an ambitious, but clear image. It aspires to a simpler technology, from an era when cities changed little from generation to generation. New urbanism looks to a time when change did not define experience, when urban traditions were sustained over generations, when individual identity seemed intrinsically linked to place.

For better or worse, most of us inhabit cities, suburbs, towns, or villages. We want to make those places good for experiencing our lives. New urbanists are trying to envision and create the good community, to enhance the normative theory of urban form. In that pursuit they join utopians and others who have preceded them in the search for the good community.

New urban approaches are already influencing popular ideas of the good community. Many municipalities have engaged in visioning processes in recent years to set the public agenda for strategic action and investment (Shipley and Newkirk 1998). Visioning exercises typically generate statements of aspiration reflecting the public desire to create communities that are healthy, livable, safe, caring, moral, and beautiful. Unfortunately, such visions often say less about particular places than they do about popular cultural values (see Table 1.2).

Within vision statements and community plans we find important cultural values. What themes dominate? We see aspirations for safe, healthy, equitable, comfortable, and productive social environments. People ask for vibrant, connected, beautiful, efficient, and green physical places. New urbanism focuses on the aesthetic questions, articulating a framework for good physical environments. In so doing, it draws the social from the physical. As Krier (1984b) argues, form cannot just follow function as the modernists would have it. Form is equally important in creating the context in which the ideal of community may be realized. 'New Urbanism seeks to create and sustain community, without seriously questioning the underpinnings or the appropriateness of this goal' (Day 2003:87). By simplifying the message, as in the minimalist vision statements so commonly found today, new urbanism ignores difficult issues of race, poverty, exclusion, and disenfranchisement in the contemporary city (D. Hall 1998). Instead it seeks consensus values within which it can operate.

Table 1.2 Sample vision statements

City	Des Moines WA, USA	Birmingham UK	Winnipeg MB, Canada
Plan	Vision and Mission Statement	Birmingham's Children and Young People's Strategic Plan 2004–2010	Plan Winnipeg: 2020 Vision
Vision statement	'A friendly and safe waterfront community'	'Birmingham with its diverse communities and neighbourhoods will be an inspiring place where all children and young people will feel secure and enjoy living, learning, developing and achieving as part of a child-friendly city'	'To be a vibrant and healthy city which places its highest priority on quality of life for all its citizens'
Source	www.desmoineswa.gov/city_gov/city_council/vision_mission?vision.html	www.bgfl.org/uploaded_documents/Childrens_ Strategic_ Plan_ 28_Sep_03.doc	Winnipeg 2001

Does the good community require a particular shape? The values in forms reveal the underlying cultural context that generates the patterns. The same form can mean different things at varying times and places. For instance, in American history the grid may have an association with democratic values and practices, but in other urban traditions it may illuminate hierarchy and function to keep people in their place (Grant 2001a). Later generations may dismiss the features we value in our communities today, just as we challenge the premises that drove the garden city planning models of our ancestors. This reality complicates efforts to identify the shape of the good community.

Many new urbanists would state emphatically that they are not trying to define a single form or set of forms as the prescription for the good community. They see new urbanism as a flexible régime that returns traditional town making principles to the table as an option for development. For Bohl (2000) the important question is whether new urbanism is better or worse at making satisfying living environments. He, and many others, would say it clearly does a better job than conventional development. In *The Lexicon of the New Urbanism*, DPZ (2003) argue that applying new urbanism principles contributes to improved quality of life for urban residents. Few people who have visited new urbanist developments come away without agreeing that these are beautiful places, beloved by their residents. If these places are not 'the form' of the good community, then they certainly provide amenable living environments.

At the same time, though, applying new urbanist principles as the formula for planning better communities remains a challenge. As Krieger (1991) notes, our dreams and values work against good communities. Our desire for privacy, individuality, elbow room, and cars make many of us resistant to new urbanist prescriptions. For every household willing to buy a home in a compact transit node, another two to four households may choose detached homes in the suburbs. For every household willing to buy a unit in a mixed income and mixed use neighbourhood, more households prefer homogeneous districts. Despite the influence of new urbanist prescriptions in plans and vision statements, those buying and selling homes in the city may not share the new urban view of what makes a good community.

Krier (1984a) talks about a charter for reconstructing the city: a moral project that mirrors a political constitution. New urbanists have adopted many charters through the years, laying out their principles for inspection and inspiration. Their ambitions are clear. For new urbanists, the form of the city is deeply connected to the fate of the city, and to the health and happiness of its people. But who will decide what makes a good city and which people will benefit from the way in which we develop? New urbanists assume that people generally will enjoy improved cities. Anyone can walk the safe and friendly streets, shop in the handy commercial districts, watch children play in the parks. The poor will no longer find themselves crowded into ghettos from which they have no hope of escape. Workers will walk or take trains to work, cutting down on transportation costs. Taxpayers will face reduced long-term costs because of the efficient urban infrastructure. Although the theorists of new urbanism insist that good urban form merely creates opportunities for social development (what they usually call 'environmental affordance'), new urbanist practitioners reveal a stubborn streak of environmental or spatial determinism. In their vision, everyone wins. Good urban form leads to a better society. If only life were like that.

The real and troubling experience of many urban environments illuminates the challenge of finding solutions to intractable problems through planning. Although we are unlikely to remedy our ills without planning, we find no guarantees that planning can solve our worst problems. Despite public celebrations of diversity, few private neighbourhoods welcome poor residents. The most successful new urbanist projects – except for a few European and Canadian examples which boast reasonable levels of public or affordable housing – have become affluent enclaves where working people cannot hope to find a home. Even as designers talk about how pedestrian-friendly and transit-oriented some developments may be, residents still lead car-oriented lives. The most beautiful new developments may meet the physical definitions of the good community, but do not address the social or sustainability objectives.

Should we expect to find the good city filled with beauty, harmony, and happy people? Marcuse (2000) reminds us that real urbanism involves dirt, disorder, congestion, and even poverty: it always has. Perhaps the search for the good community diverts our attention from dealing with specific issues we can identify as problems and tackle if we have the political will. Or perhaps the search for better design reminds us of what is important in cities over the longer term (Frey 1999).

Is the search for the good community a utopian goal, a pragmatic vision, or the holy grail of planning? Dominant paradigms implicitly promise the good community. Planning practice seeks to produce good communities. But the goal eludes us. We redefine the strategies that may lead to good communities. We contest the meanings of our visions and the causes of our failures to achieve them. The search for the good community is deeply embedded within our profession. New urbanism is but our latest method of carrying forward the search.

CHAPTER 2

GETTING TO (SUB)URBANISM

Historical context

Modern town planning had its roots in utopian communes. Through the last two centuries it gradually shifted its definitions of the problems of urban living and its options for solutions. We can better understand the role of new urbanism in the contemporary period if we have a sense of the history of ideas about good urban form. This chapter explores what planning said about urban form and community character through the late nineteenth and most of the twentieth century. In the early twentieth century, the garden city became the dominant theory of urban form in planning. Through the post-war period, the garden city concept influenced planning practice on almost every continent (Ward 1992). Planning and development practice, however, reduced the paradigm to the 'garden suburb' at best, and to an underlying ideology that engendered sprawling formless growth at worst: success robbed the garden city paradigm of its theoretical promise.

In this chapter, I briefly review the contributions of the garden city model to reveal its cross-cultural appeal and its influence over (sub)urban development. Until the late twentieth century, garden city ideas strongly influenced theories of physical planning, although practice increasingly diverged from the ideal. Through wide application of a limited number of its principles by a growing range of development interests, the original idealism of the garden city idea ultimately dissipated in practice. This analysis will set the context for examining the innovative contributions of new urbanism and the demand for compact cities and urban villages in the late twentieth century.

The success of the garden city model in appealing to professionals in so many nations indicates its ability to unite the key values of the discipline in a synthesis that promised to deliver good communities. The challenge in practice, as time proved, was to translate the rhetoric of the theory into new social and physical forms on the ground.

Theoretical models that define problems in ways that resonate with people – by connecting to popular values and with means that seem attractive – may gain general acceptance. By the late nineteenth century, that was the case for the garden city idea. By late twentieth century, new urbanism enjoyed a similar florescence.

In the early days of town planning, reformers hoped that governments that implemented rational efforts to improve urban conditions could create the good city.

Unfortunately, experience shows that many of the initiatives taken to fix nineteenth-century problems launched the trajectory that generated twentieth-century problems. We can plan for the future, but we cannot control events. Heaven only knows what issues contemporary solutions will generate for our descendants.

THE FAILURE OF THE INDUSTRIAL CITY

The ideas about good form that shaped the urban landscapes of the twentieth century reflect the failures of the nineteenth-century industrial city. Dreadful living conditions, of the sort described in a Dickens novel, began to inspire those with means or influence to seek alternatives to the crowded, unsanitary, and ugly cities they experienced. The late nineteenth century generated an optimism that reform was not only necessary but possible: that concerted societal action could create places where poverty, drunkenness, and violence would yield to good communities of happy people. That faith underscores the dominant movement of the first eight decades of the twentieth century: the garden city.

Why did the garden city model appeal so widely and have such an influence on urban form in the twentieth century? We begin with a brief review of its origin amidst a series of alternative depictions of good community design in the late nineteenth and early twentieth centuries to get a sense of why it thrived while other approaches did not. We find that the garden city succeeded in defining urban problems in a way that spoke to local community values while also offering governments strategies that met economic and organizational objectives in a broader social and political context.

The garden city model gave the incipient town planning profession a theory that incorporated the broadest range of professional values along with a clear set of principles and tools to demarcate professional expertise. Moreover, the paradigm suited the interests of an expanding market sector – the home building industry – eager to address the aspirations of a growing middle class. Governments anxious to regulate market activity to achieve political objectives (like higher home-ownership rates) transformed the principles of the garden city into the rules of urban and suburban development.

While contemporary urban form arguably has roots deep in the history of European civilizations, we will focus our history of the garden city on the nineteenth and twentieth centuries. Table 2.1 describes the key movements that contributed most to contemporary thought about good community form. Each approach defined the problems of the city slightly differently; however, the advocates of planning concurred in seeing disorder as a key issue. First and foremost, planning seemed to offer the tools to restore order in cities succumbing to the chaos of industrial capitalism. Beyond that core search for order, however, the varying planning approaches implemented different constellations of values. They suggested divergent solutions: some spatial, some social. All engaged in the search for the optimal model of the good community. Some of the options inspired experiments; some initiated movements; some sold thousands of houses; some rarely transcended the sketchbook.

MODELS FOR GOOD FORM IN THE NINETEENTH CENTURY

The early movements of the utopian socialist communities and model factory towns both promoted key social goals that sought economic adjustment and improved conditions for workers. By the early nineteenth century, industrial capitalism had generated escalating poverty and labour unrest in Britain and Europe. Low wages, shoddy housing, and chronic disease seemed to threaten the social order. While the utopian socialists took a more radical approach to solving the city's problems than did the factory owners, together they effectively set the stage for what would become the dominant design paradigms of the twentieth century.

Utopian socialists defined the problems of the industrial city in terms of the failures of capitalism to provide decent wages and living conditions for the working classes. To right the wrongs of the capitalist system, and the governments that supported it, they advocated constructing new communities according to prescribed forms. For the most part, the utopians found inspiration in socialist doctrine which proved popular among social critics in the mid-nineteenth century. They saw planned new communities as offering a strategy for recreating social order in ways that could improve the lot of working people (Benevolo 1967).

One of the most important utopians, mill owner Robert Owen, put his religious and moral beliefs into practice in New Lanark, Scotland, by improving conditions for his workers, reducing working hours, building housing, establishing educational programmes, and opening child care facilities. To better realize his ideal community form, he began to plan a new community for 1200 people on 400–600 hectares. The settlement would house dormitories for children, a public kitchen, an infirmary, gardens, and a factory. His plan showed a group of buildings organized around a large square. The community Owen founded at New Harmony, Indiana, in 1825 soon failed. Owen returned to England where he remained an active participant in the cooperative movement (Creese 1966).

The French utopian thinker, Charles Fourier, also inspired communes in North America and beyond. His 'phalanstery', a single building compound for all functions, would house a classless community of 1600 people (Vidler 1968). With complicated mathematical formulas, Fourier laid out a community form designed for architectural and social harmony. Communal services in the central area and ground floor would include library, dining halls, observatory, post office, and meeting rooms. The buildings

Table 2.1 Models of good community form

Movement	Defined problems	Dominant values	Proposed solutions	Fate of the movement
Utopian communes	poor living conditions inequality greedy capitalism	equity participation self-realization cooperation	new social order ideal community form education shared housing	• lack of funds • internal discontent • too radical • failed
Model factory towns	labour unrest poor housing immorality	hierarchy efficiency morality amenity cleanliness	cleanliness green space spaciousness good housing education	• costly to implement • labour unrest continued • limited experiments
Technical and regulatory reform	disease land conflict inefficiency	efficiency health amenity cleanliness	separate uses building codes regulations	• became strategy for bureaucracy • monotonous streets • successful
City beautiful	ugliness deteriorating city competition need for identity	amenity beauty growth competitiveness urbanity tradition	gracious buildings civic centres classical architecture boulevards	• costly • lacked popular support • scale too grand • popular for capital cities • limited experiments

Garden city	sprawl huge cities out of scale separation from nature costly housing poor conditions	equity amenity health efficiency family community nature rural life	develop on cheap land control growth limit density emulate natural patterns separate uses	• became dominant idea of twentieth century • simplified over time to mean wide lots, winding streets, parks
Neighborhood unit	car / pedestrian conflict lack of identity and community	amenity family efficiency community	separate traffic separate uses school at centre	• became very popular • fused with garden city idea
Modernist city	sprawl nostalgia hierarchy obsolescence muddle	technology equity efficiency urbanity functionalism	high rise high density towers in park land use segregation road hierarchy	• popular for city centres and public housing • some elements fused with garden city
New urbanism	sprawl car oriented development ugliness	amenity equity walkability community tradition	mixed use increase density mixed housing urban standards modified grid	• new professional ideal adopted by local governments in late twentieth century • popular with some segments of market
Urban village	lack of character, sense of place, and community lack of affordable housing	urbanity amenity equity walkability community tradition village life	vernacular/classical style mixed use increase density mixed housing village standards vernacular style	• hard to achieve affordable housing • danger of reduction to design elements

encircled open courtyards, with connecting corridors above ground level. Over forty experimental communities founded in America between 1840 and 1850 adopted Fourier's communitarian principles. Few lasted a decade.

In many ways, the utopian communes represented a fringe element in nineteenth century society. While the ideas of their founders stimulated thinking amongst the intellectual classes, the experiments did not initiate a mass movement to a new community form. The utopian mission to create a new social order via improved community form proved hopelessly radical and naïve.

Another movement from the mid-nineteenth century had a greater impact on ideas of good community form. Riots and protests drew attention to workers' discontent with working and living conditions in industrial cities. Enlightened factory owners in several nations responded to deteriorating conditions by trying to make improvements in housing as a way to increase harmony and productivity. The most famous of the American planned company towns was Pullman, Illinois. At the Prague Exposition in 1897, Pullman won the prize as the most perfect planned town in the world. Within a few decades, however, it was largely forgotten, absorbed into the growing urban fabric of Chicago, and disgraced by a labour dispute that turned violent (Buder 1967).

George Pullman built his community south of Chicago in the early 1880s, around a new rail car factory. The town was well-planned, with all the necessary amenities, except for taverns. In the American style popular at the time, it followed a grid layout. Ornamented brick buildings surrounded by flower beds created an attractive town. For its first decade, Pullman seemed the living embodiment of the good community. Visitors came from far and wide to see it, especially during the Columbian Exposition of 1893. After a nasty strike in 1894, however, the community's reputation suffered irreparably, undermining American support for planned towns (Buder 1967).

In the UK, William Lever developed the picturesque community of Port Sunlight near Liverpool in 1887, with gently curving streets following natural contours. George Cadbury soon began Bournville near Birmingham. Both these communities integrated lavish green spaces and gardens into the community form. Instead of the stark utilitarianism of the earlier factory settlements, these towns aspired to amenity and identity. Commissioned buildings of beauty and character immediately inspired emulation and imitation. Tourists flocked to visit. These communities helped to lay the ground work for the garden city movement (Cherry 1996; Harrison 1999; Hubbard and Shippobottom 1988).

As Creese (1966) argues, while the industrial community builders certainly looked after their own interests in improving conditions for their workers, they exceeded what was expected of industrialists at the time. They invested in communities in the belief that quality urban form could improve morality and productivity.

Many of them were practical Christians, committed to using their wealth in the service of the community. In Britain, they enjoyed respect for their generosity in building facilities and services for their workers. In America, they fared less well, finding themselves accused of heavy-handed tactics. Their paternalism produced good housing and clean communities. It also generally enhanced worker productivity, especially in the British examples. Although relatively few corporations followed their lead in investing so extensively in community development, these factory owners demonstrated that good planning could have salutary effects, both spatial and to a lesser extent social. Thus by the end of the nineteenth century, the examples of the model towns had set the stage for governments to consider the prospects for regulating community form through town planning.

Unlike the socially-inspired utopian and model town movements, two other late nineteenth-century movements, regulatory reform and city beautiful, focused on the efficiency and economic competitiveness of cities. They inspired government action to make cities stronger. The British and Germans took a strong position early, imposing regulations not only for sanitary reform but also for street layout. In the USA, government often encouraged market sector action, but also imposed regulations when necessary to trigger appropriate market responses.

The filth theory of disease in the nineteenth century motivated health reformers to clean up the environment in the burgeoning cities. If the accumulation of filth created health problems, activists argued, then cleaning up the city had to improve conditions. Piped water and water-carried sewer systems were installed as the century came toward its end. Cities hired engineers as technical specialists to identify and resolve sanitary problems. A growing municipal civil service saw its mission as finding ways to make cities healthier and more efficient.

Official government responses to the filth and tangle of industrial cities like Manchester and Leeds led to the British Public Health Act of 1875. The act prescribed straight, wide streets for orderliness, uniformity, beauty, and health. Later laws facilitated slum clearing. Through enforcement of the act, the 'bye-law street' became commonplace: long straight rows of housing replaced the meandering courts and cul-de-sacs of the old British cities. As Creese (1966) suggests, however, the rationality of neo-classicalism produced a cold and desolate streetscape with rows of monotonous terraced housing. The technical 'solution' would thus generate problems to which later models, such as the garden city, could respond.

The twentieth century saw technical approaches dominate the regulation of land in much of the Western world (with negative impacts that new urbanism has clearly documented). Standards for housing, parks, playgrounds, and other community facilities would increasingly homogenize urban form. All levels of government chose to apply rules and regulations to achieve desired ends, whether in urban form or in consumer behaviour. Southworth and Ben-Joseph (1995:78) suggest that visionary ideas of creating better and safer communities gradually 'evolved into a rigid, over-engineered approach'.

Out of the 1893 Columbian Exposition in Chicago came a characteristically American movement for good community form (P. Hall 1988). Daniel Burnham designed a 'White City' for the World's Fair site. With their columns and ornamentation reflecting the principles of design in classical Greece and Rome, Burnham's buildings inspired awe and admiration. Here was the city as a work of civic art, as Camillo Sitte (1965 [1889]) advocated, based on models provided by classical European and ancient cities. Visitors from far and wide took away Burnham's grand ideas of a 'city beautiful' and began advocating their application.

The Beaux Arts principles of the city beautiful movement inspired plans for several capital cities (e.g., Washington, Canberra, New Delhi), and led to plans and designs for North American cities like Chicago, Cleveland, Vancouver, and Toronto. Grand civic plazas, boulevards, and buildings proclaimed the importance of the cities. An elaborate and ornate geometry flavoured the proposed site plans. Duany et al. (2000) note that the city beautiful movement enjoyed wide popularity, especially amongst civic leaders. However, although city beautiful planning generated beautification movements across North America, its prescriptions for the city did not prove practical. The key elements of city beautiful plans focused on town centres, parks, and thoroughfares. These were, for the most part, 'monumental civic center plans' (Goodman and Freund 1968:20). They proved costly to implement, and could not always gain public support beyond a small commercial élite eager to promote their city. Nonetheless, the classic revivalist architecture associated with the city beautiful movement resulted in many impressive buildings, typically accommodating post offices, banks, or corporate headquarters. In large part, these structures, with their allusions to the grandeur of Greek and Roman cultures, created the urbane ambiance of what we see today as the traditional North American city centre.

THE TWENTIETH CENTURY MODEL: THE GARDEN CITY TAKES SHAPE

As a turn-of-the century innovation that fused British and American values and practices, and that offered a comprehensive alternative to the industrial city, the garden city movement quickly gathered strength in the first two decades of the twentieth century. Born of the creative mind of a deeply spiritual visionary, and propagated by reform-minded philanthropists and social activists, garden city ideals would have immense influence on residential environments throughout the twentieth century.

A British stenographer who emigrated to America to try farming, Ebenezer Howard proved an astute observer of the ills and opportunities of urban life in both

England and the USA (Creese 1966). In Chicago in the 1870s, Howard found inspiration in the ministry of Cora Richmond, a radical spiritualist and free thinker who had a commune at Hopedale, Massachusetts. A rapidly growing railroad hub at this time, Chicago boasted trees and gardens and the moniker, 'the garden city'. To its south, lay a model suburb designed by Frederick Law Olmsted: Riverside's meandering streets and abundant green spaces fused nature and city in a beautiful setting. The romantic suburbs of the mid-nineteenth century reflected the growing influence of the middle classes and a new residential ideal (Archer 1988).

Back in London by 1876, Howard began spreading Richmond's message of mutual aid and social reform. Working as a stenographer for Parliamentary committees, he took notes for inquiries on poverty in Ireland and debates on land reform (Creese 1966). Leaders worried that cities were growing too quickly and spreading endlessly into the countryside. London had reached 4.5 million people, extensively segregated by class. A home with a garden, a common aspiration of Victorian society, proved beyond the means of most, including Howard (Fishman 1977; P. Hall 1988).

A man of his times, Howard shared a passion for the Victorian 'cult of domesticity'. He saw the home as central to family welfare and personal character development. He thought the lower middle classes should be able to aspire to such an ideal. These values played a major role in the movement he founded. Howard's ideas also reflect the common belief of the Victorian age in social reform through physical improvement. Radical physical and social restructuring would, people of the era believed, generate evolution and progress (Fishman 1977). A cooperative commonwealth of brotherhood and equality would result. Although Howard joined the Socialist League in his search for solutions to the problems of industrial society, he became distressed by the revolutionary talk (Miller 1989). He hoped to find a peaceful path to reform: community design as a means to a just social order.

In creating a synthetic option for the good community, Howard revealed his peaceful optimism (Howard 1985 [1902]). Influenced by Robert Owen's communitarian cooperative ideas and Peter Kropotkin's anarchism, Howard sketched plans for a complete and self-reliant community where residents could enjoy an independence denied in modern industrial society (Fishman 1977). Howard held that a good environment benefits everyone. He advocated development in harmony with nature, at reasonable densities, with decent housing available for all. In his early speeches and book, he framed land reform as an important element in the concept: nationalization of land would socialize the benefits of development and keep housing affordable (Creese 1966). This key component of his theory would prove impossible in practice in the Western world, and would contribute to the unravelling of some of the fundamental concepts of the garden city paradigm.

Howard also argued for the benefits of a system of cities, with towns of about

30,000 people in 400 ha being ideal (and another 2000 people in the agricultural belt of just over 2000 ha). He sought ways to prevent the sprawl of London from continuing to consume productive agricultural land. His illustrations show a central city (of 58,000) with six satellite cities equidistant around it, each housing its own industry and services, and reasonably self-sufficient in meeting the needs of its population. Belts of agricultural land separate the cities. The federated social city would include about 250,000 people in total (Miller 1989).

The garden city concept as Howard illustrated it would have at its centre a park with civic facilities in the middle, surrounded by residential districts linked to work sites by broad boulevards and transportation systems (Howard 1985). Rail, canal, and road networks would link the cities. Houses should sit on good sized lots, with ample green space and urban amenities. The city would be clean, efficient, manageable, and affordable. The residential precincts would bring town and country together in a perfect combination: as Chiras and Wann (2003:4) put it: 'The ideal of the suburb was country homes for city people – nature without the mud'.

Howard's ideas resonated immediately with key segments of Western society. Both British and American culture revered the rural life, and saw in nature the possibility of redemption and health. They also recognized that the city could offer benefits not available in rural communities. Thus a model of urban development that sought harmony between town and country – that offered the best of both – had immense popular appeal. In Britain, a Garden Cities Association formed quickly and began work to build a model garden city. Similar organizations grew in other countries as well. The first garden city new town in the UK, Letchworth Garden City, soon revealed the challenges of articulating Howard's vision of the good community. Instead of building on public land, the Association had to raise funds to buy land. Costs escalated. Howard had hoped to keep rents low, but in practice that proved challenging. While Raymond Unwin and Barry Parker designed a community that has stood the test of time in terms of its beauty, greenery, and elegance, Letchworth met few of the visionary principles that motivated Howard (Miller 1989).

Parker and Unwin had a considerable impact on design traditions of the twentieth century (Creese 1966, 1967). Advocates of the Queen Anne revival style promoted in the arts and crafts movement, they often designed suburbs that evoked the image of the traditional English village. The design for Letchworth reveals the influence of city beautiful thinking, as well as garden city ideas. Since Howard had not prescribed a road layout in his concept, Unwin and Parker could select attractive models from other paradigms. An elaborate geometry, clearly inspired by city beautiful notions, occupies the centre of the town, with a wide tree-lined avenue leading to the core. A 'ring road' loops around the modified grid core. Part of the road layout reminds me of an insect's body, with diagonal roads radiating from the 'head'-like antennae. The elegant layout would inspire many imitators.

By the time they began their next major project at Hampstead Heath for the National Trust in 1905, Unwin and Parker were already modifying the model of the garden city. With each iteration, it lost some of its original features and adapted to market realities (see Figure 2.1). Hampstead Garden Suburb was by no means a complete community. A middle class suburb with rail connections to London, Hampstead embodied the design principles that Unwin and Parker were establishing, but lacked the social commitment that characterized the garden city movement. The design constituted a major break with nineteenth-century building patterns. Parker and Unwin had a private member's bill passed in Parliament to allow them to avoid the straight and endless 'bye-law streets', which they felt limited design options. Henceforth, gently curving streets and cul-de-sacs became their hallmarks. The layout protected a large area of heath land. Radiating diagonals are interspersed with curving lanes that follow the topography of the site, creating a hierarchy of streets. Each house had both a street and a garden view.

We see, then, within only a few years of the garden city theory coming to the fore to inspire the new town planning movement, its key principles were already diluted in practice, subtly undermined by the garden suburb idea. Social vision quickly yielded to design and planning principles that appealed to the market. The success of Hampstead Garden Suburb created controversy in the garden city movement. Those who advocated satellite garden cities felt that using the new principles for suburbs defeated the purpose of the movement. Others applauded Parker and Unwin for the beauty of suburbs and for proving the widespread appeal of planning principles. Their incremental approach allowed designers and planners to

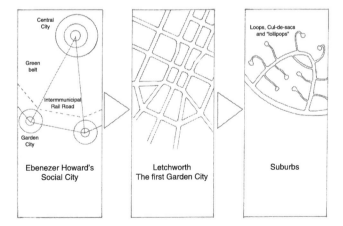

Figure 2.1 The garden city reduced
The garden city (panel 1) offered a conceptual model of the social city: equitable, connected, contained. In building Letchworth (panel 2), the designers compromised on the social objectives. (Only the town centre is shown.) Through time, the concept was reduced to isolated pods of cul-de-sacs and loops (panel 3).

apply new principles even in smaller development projects (which were easier to finance than new satellite cities). The pre-eminence of the garden suburb in the twentieth century reveals the impact of Unwin and Parker's contribution and the easier integration of a partial notion of the garden city than the full-blown model (Ward 1992).

The garden city concept travelled far and wide during the period from 1900 to mid-century (Ward 1992). For instance, Watanabe (1992) indicates that the Japanese interpreted the garden city in a way that fit contemporary priorities. Translating the term as *den-en toshi* (literally, countryside city), Japanese planners and developers began to build garden suburbs in the countryside, linked by train to growing cities. Reformist notions disappeared from the model, as did the idea of self-sufficient communities. Instead, the traditional Japanese glorification of country life became a way of selling villas to a new upper-middle class. Suburban development fragmented rural landscapes in an inefficient manner. Watanabe argues that instead of limiting sprawl, the Japanese garden city exacerbated it. The British concept of a complete satellite city never took root in Japan, although it did travel to other countries like Sweden and France.

Through the twentieth century, elements from other theories were added to garden city principles as the model was popularized and transformed. The 'neighborhood unit' theory of Clarence Perry (1974 [1929]) created an important concept for the design of districts within the city. Perry argued for residential areas anchored by a school and other community functions at the centre. A hierarchy of streets, as implemented at Radburn by Clarence Stein and Henry Wright, would protect residents within the neighbourhood from the growing menace of automobile traffic (Birch 1980; Buder 1990).

The modernist city models of Le Corbusier inspired a lot of interest, but few city plans (Fishman 1977). In the Contemporary City and Radiant City models of the 1920s and 1930s, towers reach to the sky to create a compact and thoroughly modern city. Abundant open space at ground level provides lots of room for people to enjoy air and light in an urban context. While few completely modernist cities were built, though, some of the ideas of the modernist movement became fused with garden city ideas to generate the dominant planning principles of the twentieth century (Barnett 1986). High-rise towers for commercial development and for affordable public housing became common planning and design solutions in the 1950s through 1970s. 'Towers in the park' seemed a way to bring natural elements into an urban context while using sites efficiently and economically (Relph 1987).

THE FATE OF THE GARDEN CITY

Why did the garden city, fused with neighbourhood unit theory and an admixture of modernism, become dominant in its time? How did it mutate from the idealism of Howard into the framework of suburbia? The twentieth century was an era of substantial urban development and redevelopment. As governments sought an interventionist role in monitoring and managing growth, they needed theories and prescriptions to legitimate their actions. Garden city ideas provided a set of beliefs and practices that justified particular policies and regulations in the urban landscape. As a flexible amalgam of concepts that ranged from the grand idea of designing complete new communities to simple notions about attractive street patterns, this popularized garden city theory could appeal to a wide range of interests and values. For those concerned about equity, it promised affordable housing and well-serviced communities. For those committed to amenity, it boasted ample green space and attractive civic facilities. It ensured healthy cities by pre-planning adequate infrastructure. It promised more efficient use of community resources. It created urban spaces designed to meet the needs of families in landscapes that retained a place for nature. It seemed a perfect theory for a time of growth and optimism in the future.

In the UK, garden city theory not only inspired community-based movements, but generated government action as well. Interest in the garden city idea set initiatives in motion that would eventually make garden city theory government policy. After the Second World War, the British government implemented new town legislation and built a series of communities according to modified garden city ideas (F. Schaffer 1970). Designed to take growth away from London and some of the other big cities, satellite cities would manage the threat of endless urban expansion while creating desperately needed good quality affordable housing.

Thus British practice illustrates the implementation of many of Howard's original garden city ideas: green belts to limit urban growth, complete new communities with a mix of uses, adequate social housing on public land. Of course, these new towns are hardly the social cities that Howard envisaged. Through time they became increasingly large, exceeding the scale he had proposed by a factor of more than ten. As government-run and organized settlements, these places lacked the independence that Howard advocated. They had some of the design features promised by the garden city movement, but none of the sociological underpinnings of Howard's social city.

In the 1920s, the Regional Planning Association in the USA sought to apply garden city ideas to community planning. The model community of Radburn was to implement the concept. It fused some of Unwin and Parker's design ideas from their garden suburbs (like cul-de-sacs and large open areas) with new concepts (like street hierarchies) intended to respond to the menace of automobile traffic. Perry's neighbourhood unit derives from concepts tested at Radburn. Unfortunately, the Depression ended the Radburn experiment, and new town projects proposed during the 1930s as part of Roosevelt's New Deal could not save the concept of government-initiated community building (Birch 1980; Buder 1990; D. Schaffer 1982).

While the idea of building complete new towns was left to the private sector in America, many popularized garden city ideas became embodied in the principles and practices of North American zoning and land use regulation. Segregated land uses, density limits, green space requirements, design quality requirements, and separation of pedestrian and vehicular traffic all find their roots in ideas associated with the garden city model of urban form. Physical elements which proved attractive for marketing suburban development became entrenched in planning and design practice. In the USA, even more than in the UK, the social objectives of Howard's garden city theory disappeared even as the design elements associated with it became omnipresent.

By mid-twentieth century, suburbs generally proved economically and demographically homogeneous, with single detached houses and institutional uses dominating. Suburban living dominated mass culture in the age of television. Two-parent families with young children constituted the most typical households buying in the suburbs. Higher density residential and rental properties were kept apart. Planners placed new commercial zones on the periphery of the city, along major arterial roads: strip development of new retail chains resulted. The ideology of the post-1945 city venerated the separation of uses, and idealized the private automobile as an agent of freedom, mobility, and choice.

Thus we see that the values that supported post-war suburban communities included:

- desire for the best of country and of town (low density housing in an estate-like setting);
- belief in the need to protect the domestic realm (women and children) from the dangers of the city;
- faith in progress, science, and technology;
- wish for individual ownership and privacy.

The mid-twentieth century suburb generated an urban form that celebrated new modern lifestyles. It suited a time of economic expansion, when a broadening base of the population came to share growing industrial wealth. The working and middle classes finally had the means to afford decent housing, and even to become home owners. Demographic and social change transformed family lifestyles, giving households new options and mobility. In the same period, an expanding home-building and development industry needed to market its product, and took advantage of a mass media attuned to appealing to societal values (Ward 1998). Relatively low land and transportation costs made suburban land affordable and desirable. Governments at all levels promoted development and extended their roles in managing growth. It proved feasible to 'blend town and country' through building houses on large lots in the urban fringe. These factors in concert created the cultural context in which the suburbs increasingly came to reflect popularized garden city design principles. As the design features dominated practice, however, the social premises behind the original theory slipped farther away. Howard would not recognize his garden city ideas in the suburbs of the late twentieth century, even though his critics may blame him for contemporary problems.

The garden city model sought to reconcile nature and the city in residential environments. At the time it originated, the middle classes were exercising their political might in cities struggling with the consequences of industrialization. The idea of integrating green areas into the city appealed to those for whom rural landscapes represented the epitome of beauty and respite. While the theory promised an effective synthesis of nature and culture, a century of popularized garden city practice resulted in suburbs that many would argue are neither natural landscapes nor positive cultural symbols.

In some ways, we might argue that the landscapes that resulted from the garden city model reflected religious values that assumed human domination of the Earth and economic values that supported mass consumption and the commodification of housing. The wide open front yards of the garden city suburbs represented icons of success for the growing middle classes: the conspicuous consumption of space for display purposes. As farm practices changed, farmland near cities became increasingly available for suburban expansion. Government policy and lending practices facilitated suburbanization. Cheap mass-produced cars and gasoline fuelled both economic and urban expansion. The garden city had democratic appeal. Conformity in design was accepted, and perhaps even coveted, as Levittown proved (Kelly 1993). The garden city provided an ideal landscape for the nuclear family at a time of family growth, and helped to support an expanding capitalist economy into the 1970s. It fused well with the model of progress that powered modernism, and reflected hope in a better future.

Of course, the garden city's theory of self-contained communities – with a balance of jobs and housing – was seldom realized, except to a limited extent by socialist governments creating new towns, and with a few master-planned private towns in the USA. While we see that values of health, amenity, and efficiency remained strong ideals throughout the implementation and popularization of the garden city model, the value of equity got short shrift. Only socialist governments that built new towns have a reasonable history of providing affordable housing in garden cities. More commonly, the premises of separated uses, street hierarchies, and open landscapes proved quite successful: this model of good urban form supported property values in a vibrant housing market.

Through time the garden city model was adapted and popularized differently in local contexts. Its users picked the elements that worked for them. In North America, garden city theory was eventually reduced to winding streets, detached houses, and wide open lawns (Grant 2002a). In the UK it led to leafy crescents full of semi-detached houses, terraces of council homes, and open grounds featuring tower blocks. In Japan, it caught on more in the title of new developments than in the practice of community design.

Curiously enough, even as the garden city stands challenged for environmental irresponsibility and social fragmentation by contemporary urbanists, garden city premises still appeal to many home purchasers. Couples looking for sympathetic environments in which to raise their families continue to look to the suburbs as a principal choice. The garden city model captured values and aspirations that have not disappeared through a century of practice. Perhaps we should see the new suburban landscapes of snout-garage neighbourhoods as contemporary extensions of the garden city model.

The modernists who influenced development through the twentieth century hoped to empower the disadvantaged and reduce the influence and inequities of capitalist culture (Harvey 1989). They wanted to promote equity, and to bring air and light into the city. They supported urban renewal and public housing to promote improvement in the inner cities. Despite their extensive influence over building practices in the twentieth century, they did not achieve their aims. Instead, modernist design became inextricably associated with corporate culture, disregard for the disadvantaged, and destruction of quality urban environments.

The failure of the modified garden city model and of the modernism linked with it reveals the challenge of planning the good community, especially in countries with weak planning controls. A phenomenal cultural consensus about the goals of urban development and the means required to transform society in desired directions could not effect the desired changes except in those nations where government provided political will. Indeed, the application of garden city principles and modernist building technologies generated townscapes in much of North America that by the end of the twentieth century seemed seriously flawed. Cities sprawled endlessly into their peripheries, wasting land at a voracious pace. Too many twentieth-century buildings eschewed beauty for monotony. The inhabitants of the increasingly ubiquitous suburbs found themselves ensnared in frightening traffic jams. Governments sank into debt to build the roads, service the parks, and transport the students the suburbs generated. Despite its mass appeal as theory, as practised the garden city had become seriously dysfunctional in many places. The dystopia of the suburbs opened space for an alternative paradigm that promised a new formula to produce the good community.

CHAPTER 3

THEORY IN NEW URBANISM

THE NEW VIEW

New urbanism takes the failures of the garden city and of modernism as its starting point. What theoretical premises and assumptions drive new urban approaches? What should we expect from new urbanism in practice? In this chapter, I briefly explore elements of the theory of new urbanism and related new urban approaches. Although the most vociferous proponents of new urbanism downplay the importance of theory while they focus on the practical applications of their ideas, we find that particular premises, assumptions, and hypotheses about the effects interventions may have do inform the movement. As in any movement still taking shape, the theories of new urbanism are multiple, fluid, partial, and sometimes even internally contradictory.

New urbanism involves a set of approaches that, while they may not be entirely congruent, share some commonalities in their various applications. New approaches to urbanism appreciate the traditional compact urban form of cities of previous centuries. By and large, they recognize the inefficiency of twentieth-century suburban patterns. They advocate redeveloping older areas of the city, especially those places abandoned by industry or blighted by concentrated poverty. They promote mixed uses at fine grain and quality urban design. Often, but not universally, they look to historical precedents for standards and advocate connected street patterns.

In its various manifestations, new urbanism reveals a strong need for order. Many authors, especially Krier (1984a, 1998), Kunstler (1993, 1996), and Duany *et al.* (2000), reveal their disdain for the visual confusion of the contemporary city. They prefer the coherent organization of the traditional city, where builders chose from a limited repertoire of designs and materials. Design guidelines and prescriptions, they suggest, will keep things in their place in the city through a unified physical vision. As Kelbaugh puts it:

> What *is* new about the New Urbanism is its totality. It attempts to promote a sort of unified design theory for an entire region – from the small scale (building block, street) through the intermediate scale (corridor, neighborhood, district) to the large scale (regional infrastructure and ecology).
>
> (Kelbaugh 1997: 132)

I find it hard to avoid concluding that new urbanism seeks to categorize things as either black or white, with few shades of grey allowed. The hated suburbs are neither town nor country: they create zones that have no clear identity in a schema that seeks to classify places as either urban or non-urban. New urbanism hopes to create places immediately recognizable as urban in their design and use. The traditional is preferred over the modern; low-rise (pre-industrial urban) over high-rise (modernist urban). A strong need for conceptual clarity and distinctiveness runs through new urban approaches. The value of relativism and mixing of styles associated with post-modern approaches receive no respect from new urbanists who believe they have identified the correct order to produce better communities (Beauregard 2002; Talen and Ellis 2002).

One dualism that appears often in new urbanist discourse involves nature and culture. A few new urbanists, like Peter Calthorpe (1993, 1994), seek to reconcile nature and culture. Calthorpe often discusses how cities can avoid threatening nature: like other new urbanists, he sees nature as separate from culture and requiring protection from a system with different values and processes. Many new urbanists, however, seek to keep nature and culture in their proper places. For example, Duany et al. (2000) and Duany and Talen (2002a) suggest the city should avoid including inappropriate natural elements: the city requires formal treatment (as in neat rows of trees), not natural stands of wild plants. Scully (1994:221) takes the argument to its logical conclusion when he argues that architecture creates 'its own model of reality within nature's implacable order. It is within that model that human beings live; they need it badly, and if it breaks down they may well become insane'. In other words, architecture and design provide a mediating function for humanity. In this model, urban structure is key to sanity and survival. I will return to this theme – which privileges the designer with awesome responsibility for the well-being of society – later in the chapter.

As we look at the new urbanist conceptualization of the relationship between nature and culture we find a dramatic contrast with another contemporary planning and design approach. Landscape ecology has established a substantial following, especially in landscape architecture and environmental planning. Its advocates integrate an understanding of nature and culture as processes operating simultaneously on humans and landscapes (Dramstad et al. 1996; Forman and Godron 1986). Where landscape ecology seeks to retain natural landscape functions even in the urban context, new urbanism chooses to separate urban and natural functions in discrete realms.

The interest of new urbanism in traditional urban styles may be a product of the age. Fiscal uncertainty, moral ambiguity, and cultural diversity lead some to look to the past for simpler times. Baum (1999:7) worries that new urbanism 'may try to project an imaginary past on to the future through wishful thinking, rather than creating a workable future through strategic action'. In many ways, new urban ideas react to the problems of the post-war city: the precepts that guided post-war development thus merit intense criticism. Looking for better alternatives leads the new urbanists to search for what made traditional communities like Nordnes in Norway seem special (see Box 3.1). In the process, they make explicit their belief that good design creates good communities, and that the best places (most interesting, active, and diverse) are well-designed. At the same time, though, they cannot entirely escape the modernism they despise.

THE CALL FOR A NEW APPROACH

The new urbanists were not the first to notice that contemporary cities have significant problems, but they have arguably been among the most vociferous in their critiques. Endless sprawl, traffic congestion, racial tension, economic stagnation, and deteriorating infrastructure were only the most obvious problems of the twentieth-century city. Rather than ushering in an era of equality and comfort, though, modernist design contributed to creating monotonous and ugly cities plagued by inequity and dysfunction. Scully (1994:223) says that: 'The International Style built many beautiful buildings, but its urbanistic theory and practice destroyed the city', creating 'landscapes of hell'. Many new urbanists believe that modernist design worked against place making, and significantly undermined urban quality (Cusumano 2002).

The efforts of garden city advocates to build new towns met with some success, especially in the UK and Europe, but for the most part they could not resolve the urban problems of the late industrial age. Indeed, Calthorpe (1994:xv) lambasts Ebenezer Howard's movement as 'a Luddite's vision', unsuited to the task at hand. Instead of building stronger and better cities, the garden city movement led to suburban expansion, and what Krieger (1991) calls parasitic garden suburbs.

The rational planning activities of the twentieth century sought to order the city: to make it efficient and functionally coherent. In the process, planning contributed to new problems such as traffic congestion and poor quality urban design. As Bressi (1994) notes, planning set up a regulatory framework but then left design to developers. That system created separated uses and uniformly poor design. The resulting suburbs had few redeeming values for the new urbanists. 'Most post-World War II suburbs were bedroom communities. They were great places to raise families, offering good housing at reasonable cost. But they weren't true places. This, of course, is the crux of our suburban dilemma today' (Cusumano 2002:ix).

Krieger (1991:15) notes that for new urbanists like Duany and Plater-Zyberk

Box 3.1 Strong traditional neighbourhoods: Nordnes

Nordnes is a neighbourhood in Bergen, Norway, located on a peninsula at the edge of the city. Historically, the neighbourhood housed the city's labourers and poor. It was also home to one of the largest wooden ship-building wharves on the west coast of Norway. The neighbourhood was never planned, but grew organically over a period of 300 years.

In the 1970s, city officials proposed to demolish parts of Nordnes. At the time, the arguments for renewing older neighbourhoods were common among politicians, architects, and planners in European cities: cleaning up blight. In reaction to the proposed demolition, many students and recent graduates living in the city protested. Some of these students lived in Nordnes at the time and were to be displaced; other families had just bought the relatively cheap housing as their first homes. After a long battle, the protesters managed to save the neighbourhood.

Over the part 30 years, Nordnes has undergone what many planners might call a 'gentle urban renewal': a kind of urban repair. Residents in Nordnes formed

Figure 3.1 A traditional street in Nordnes
Residents have improved older homes and adapted effectively to life with few cars in this traditional neighbourhood in Bergen, Norway.

committees to restore the dilapidated wooden houses, then to build a series of small parks, and plant street trees. Now they are making a network of pedestrian streets and car-free zones. The young urban pioneers have raised their families in Nordnes and the neighbourhood is well-established in the city fabric (see Figure 3.1).

Many of the residents feel an intimate relationship with the neighbourhood that they saved. A short walk around Nordnes reveals a community rebuilt by hand in all its detail. The people living in Nordnes are active citizens. The neighbourhood has its own newspaper and web page. Festivals animate the streets several times a year. Small businesses are continually opening, and the neighbourhood seems lively and thriving. Today, Nordnes is a vibrant neighbourhood.

From the perspective of someone who has been involved in designing new neighbourhoods, I see in Nordnes features that make it different from planned neighbourhoods. The scale of the streets and houses relates to humans. The streets were never planned for cars. There are few building standards, allowing owners to express themselves in their homes: this creates heterogeneous architecture. One can walk through Nordnes every day for years, and still discover new details and spaces. A neighbourhood based upon standard regulations, no matter how progressive, could never match the diversity found in Nordnes.

Time is an element that we often forget as architects and planners. In a rush to make the so-called 'perfect neighbourhood', ambitious architects and planners try to design qualities that can only be generated over time. Nordnes has grown slowly, giving the neighbourhood depth, allowing the citizens to build connections, and giving them time to correct mistakes in each successive stage of development. How could we plan that?

Source: Todd Saunders

'the modern suburb [is] a rudimentary form of habitation, something . . . in need of civilizing'. *The Lexicon of New Urbanism* defines 'suburban' as meaning less than urban (DPZ 2003). Krier (1984a) argues that the suburbs are anti-city: they kill both the city and the landscape.[1] Marred by car-dominated streetscapes, they lack the beauty and community of urban neighbourhoods of the small town or of genteel late-nineteenth-century districts. By 1990, American suburbs had more than half the people and half the jobs in the nation (Southworth and Owens 1993). The new urbanists want to turn the situation around. They envision cities that lack suburbs, but instead consist of multi-functional and interconnected urban quarters: places to live, work, shop, and play.

New urbanism has urged all-out war on sprawl and car-dependent development forms. Its proponents have criticized the patterns in which cities grow endlessly into fields, sprouting at the end of highway off-ramps (Mitchell 2001). New urbanists decry cities that seem designed for cars instead of people. They challenge the oversized streets and parking areas in shopping districts, arguing that pedestrian-friendly environments are preferable. They criticize houses with large garages facing suburban streets. They want to bring jobs and residents back to urban districts to end what Cervero (1986) called suburban gridlock.

At the root of the problems of the city are the rules used to manage land. 'Our planning tools – notably our zoning ordinances – facilitate segmented, decentralized suburban growth while actually making it impossible to incorporate qualities that we associate with towns' (Krieger 1991:9). Krier (1984a) says that zoning rules destroy the social and physical fabric of communities. Kelbaugh (1997:118) criticizes the 'visual messiness' that results. While controlling uses, zoning allows a mishmash of styles and building types that creates visual incoherence instead of harmony. Salingaros (2004:online) quotes Demetri Porphyrios as saying that Krier recognized 'that indiscriminate toleration posing as the guarantor of democratic freedom had thrown architecture and the city into disarray'. Zoning rules did not guarantee urban order. Too much choice creates chaos.

According to new urbanism, planning policy in the post-war period played a major role in contributing to sprawl, undermining the public realm, and diminishing civility. Heavy-handed governments and lenders, and out-dated zoning regulations, encouraged wasteful development patterns (Zelenak 2000). Planning and zoning segregated and concentrated the poor, and resulted in standardized ugliness. Zoning is 'overly verbal and complicated', say Moule and Polyzoides (1994:xxi): 'Zoning conflates issues of use, density and form to such an extent that it has spawned the unpredictability and visual chaos typical of the American city.' Only an approach rooted in design and with appropriate visual principles could offer a viable option.

In recent years the critiques of sprawl and zoning have gone farther than imagined reasonable a decade ago. Duany *et al.* (2000) attribute youth violence, suicides, and road rage to suburban sprawl. Kunstler (1999) suggests that school killings like those in Columbine, Colorado, reflect the social anomie of the suburbs. Lennard and Riley (2004) liken dysfunctional cities to dysfunctional families that may harm children. Since health studies have shown a higher incidence of obesity in suburban areas than in urban areas, several sources argue that the suburbs may make people fat, or at the very least that they create conditions that do not encourage enough walking (Frumkin *et al.* 2004; Kreyling 2001; Stein 2003). Improving social conditions calls for radical changes in urban form.

For new urbanism, design is the problem and the solution. Bad design (caused by inadequate planning and useless zoning) generates the inadequacies and malignancies of the contemporary city. While social critics may say that cultural values and behaviours keep people glued to their televisions, drive people to their cars, encourage racial segregation, and reward bad behaviour, the new urbanists find a simple explanation. If we find the right design formula for the city, we might create the conditions for building better communities.

THE ORIGINS OF NEW URBANISM

We can see the origins of new urbanist theory in certain ideas about the city in the late nineteenth and early twentieth century. One source commonly cited by new urbanists is Camillo Sitte, whose 1889 book, *City Planning According to Artistic Principles*, advocated applying better design principles to urban planning. Sitte rejected the long, straight boulevards appearing in European cities, and promoted the principles that shaped medieval cities. Many of his ideas stimulated the designs of garden cities, including his penchant for curving street alignments and T-intersections.

As is common in planning theory and practice generally, the new urbanists pick and choose elements of theory from diverse sources. Thus they can claim both Camillo Sitte and Raymond Unwin as inspirations while simultaneously dismissing many of the ideas for which Sitte and Unwin became famous.

The city beautiful movement of the turn of the twentieth century clearly inspired new urbanism. A design practice that focused on civic centres, classical architecture, and monumental urban design embodied many of the principles that new urbanism has brought back into vogue. The grand buildings of the Progressive Era offer North Americans some of the best examples of high quality classical urban design.

Much as the new urbanists love to hate the garden city and modernism, we see that new urbanism owes much of its theoretical development to those influences. In large part, new urbanism defines itself in opposition to these earlier movements. Where modernism placed towers in the park, new urbanism puts cosy townhouses on small lots with back alleys or courtyards. Where Raymond Unwin (1912) wrote 'Nothing gained by overcrowding' to impose a limit on residential densities, the new urbanists advocate setting minimum density targets. Where the modernists proposed technological innovations in construction, the new urbanists prefer heritage conservation and traditional architectural styles and materials. Where the garden city liked winding streets and integrated green spaces, North American new urbanism prefers a grid and formal greens. While some of the ideas that drove the modern town planning movement – like the neighbourhood with its clear edges and centre – remained as new urbanism took form, new urbanism challenged many twentieth-century conventions of urban development.

Arguably the most important initial theoretical contribution to new urbanism came not from hot-shot young designers but from a middle-aged journalist. Jane Jacobs's (1961) book, *Death and Life of Great American Cities*, constituted an acerbic critique of the urban forms generated by modern town planning and a call for more responsible ways of building cities. Jacobs argued that urban vitality comes from density, mixed use at a fine grain, and diverse neighbourhoods. She demanded an end to large parks that created venues for criminal activity and to high rise towers that warehoused the poor. She wrote about 'eyes on the street' and neighbourhoods rendered civilized by residents willing to intervene. The busy, caring streets she described resonated with a generation living in suburbs where they barely knew their neighbours.

Building on increasingly common trends in the 1970s to promote compact urban form, heritage restoration, urban revitalization, and mixed use zoning, the new urbanism articulated a coherent theory of an alternative model of the city. Early revitalization efforts had tended to produce stand-alone projects. For instance, many cities had invested in large-scale downtown redevelopment projects that featured mixed-use buildings (Collier 1974), but these were often inwardly-focused and poorly linked to the rest of the urban fabric. New urbanism offered quite a different way of thinking about the city, and of connecting its parts. Above all, the new urbanism refocused attention on physical planning and urban design. It drew on the work of Christopher Alexander (Alexander *et al.* 1977) whose search for universal principles of good form led to the development of 'a pattern language'. It built on the contribution of designers like Kevin Lynch (1981) whose theory of urban form stressed the importance of legibility and sense of place in good urban design.

The direct line of new urbanism theory begins with the teachings of Leon Krier, and his influence over Andres Duany and Elizabeth Plater-Zyberk in the late 1970s. Krier, his brother Rob, and other Europeans like Maurice Culot, followed the architect Aldo Rossi in the rational or neo-rational school of architecture. As Krier wrote in 1978, this movement sought to reinforce the rationalism of the Enlightenment (rather than the rationalism of contemporary capitalism). Using Marxist terminology in advocating a return to traditional building practices, Krier (1978:42) argued that dynamic urban culture offered greater promise of democracy and socialism than did the conservative suburbs. The principles of rational architecture included seeing historical centres as models of collective life, using urban space as an organizing element in urban form, studying typology and morphology to set precedents, recreating a public realm, and reconstructing the basic elements of street, square, and *quartier* (quarter).

Krier (1984a) attacks the chaos of the modernist city and advocates reviving classical principles. He would build walkable urban quarters with clear edges and centres, and coherent architectural styles: his emphasis on design and density differentiates his quarter from the earlier neighbourhood unit. The urban quarter 'integrates work, culture, leisure and residence into a dense urban environment, into a city within a city' (Krier 1978:163). Krier favours what he calls an authentic urbanism, grounded in the traditional principles of the European city. His concern about the loss of authenticity in urban form presages a common theme in new urbanist discourse, especially in the writings of Duany and Plater-Zyberk (e.g., Duany et al. 2000): a close reading of *Suburban Nation* reveals the preoccupation with 'true', 'real', and 'authentic' neighbourhoods and places throughout the text.

For Krier, streets are important public spaces, functioning as part of both the public and private realm. Krier (1984a) prefers an urban form with small blocks, well-defined streets and attractive squares. His view of the city appealed to an important patron: Charles, Prince of Wales. In his influential book and television special, *A Vision of Britain*, the Prince said: 'A community spirit is born far more easily in a well-formed square or courtyard than in a random scattering of developers' plots' (Prince of Wales 1989:87). The kinds of spaces that make historic European cities attractive become the design formula for meaningful new urban development (see Figure 3.2): places designed to generate community.

The American adaptation of Krier's thinking led Duany and Plater-Zyberk to look to the roots of the American small town for a model of traditional urban development. After hearing Krier speak, Duany and Plater-Zyberk set up their own architecture and planning firm (DPZ) to pursue a new approach to housing development. They studied traditional urban patterns to uncover the algorithms to reproduce in new developments. In their first major experiment at Seaside, Florida, they practised what they would come to preach (see Chapter 4).

As Duany and Plater-Zyberk were developing neo-traditional town planning, California architect Peter Calthorpe was considering the implications of sustainable development on urban form. His interest in sustainable development and healthy

Figure 3.2 The European city square
The well proportioned city square provides a container for vibrant urbanism. The Grand Place in Brussels entertains tourists and workers enjoying their lunchtime break.

communities led him to strategies for urban design that bore many similarities to those of Duany and Plater-Zyberk. Calthorpe worked on designing walkable neighbourhoods that he and his colleagues called 'pedestrian pockets' (Kelbaugh 1989). He articulated a framework that set the public transportation or transit system as the backbone of a regional settlement system (Calthorpe 1993; Calthorpe and Fulton 2001). Calthorpe recognized that human needs do not change with technology, and that 'certain qualities of culture and community are timeless' (1993:16). The two approaches shared basic premises, but employed different implementation strategies.

NEW URBANISM GETS NOTICED

In a culture hyped on the cult of celebrity (Lasch 1979), how did new urbanism make it to the news magazines (e.g., Anderson 1991; *Newsweek* 1995) and television talk shows? As Kelbaugh (1997) admits, it takes bombast and ego to get noticed today. It also helps to have rich and powerful patrons, and a product that makes good 'wallpaper' for television and visual media.

With his noteworthy head of hair and cravat tied neatly at the neck, Krier cuts a debonair figure. Although he did not build many structures early in his career, his design ideas managed to captivate the Prince of Wales. Educated in history, the heir to the British throne had a deep interest in urban form and found inspiration in Krier's approach to the city. With the resources to establish educational charities and experiment with building projects, the Prince helped Krier and other urbanists to demonstrate their vision.

Duany and Plater-Zyberk also had influential patrons to allow them to build their projects. Robert Davis gave them the opportunity to create Seaside, the birthplace of American new urbanism. Galen Weston provided the funds to develop Windsor, Florida, and to arrange the initial meetings of the Congress for the New Urbanism (Kelbaugh 1997; Thompson-Fawcett 2003b). Duany became the main media spokesperson for the movement. As Ernest Alexander (2003:online) put it, Duany is master of using 'image and rhetoric in the planning process' to mobilize commitment. Duany spoke at conferences, gave workshops to planners and developers, and offered soundbites to countless reporters. Those ready to dismiss the suburbs lapped it up (e.g., Adler 1994).

Journalists like James Howard Kunstler (1993, 1996) and consultants like Peter Katz (1994) also played major roles in promoting new urbanism. Katz told an interviewer (O'Keefe 2002) that new urbanism had been successful in marketing 'the brand' but that next the movement needs to set clear standards and specifications to ward off those who might claim to be new urbanists without following the necessary precepts. Success breeds emulation without sufficient discipline.

The Congress for the New Urbanism, created to support the movement, has become a key venue for developing ideas and strategies for marketing new urbanism. With books, reports, newsletters, web sites, and copyrighted codes, new urbanism consumes a great deal of paper, and receives abundant attention. In North America, it has drawn support from professional planners' organizations and from developers' groups. Correlate groups in England and Europe, like the Urban Villages Forum, have promoted new urban approaches there as well (see Chapter 5).

THE PRINCIPLES OF NEW URBANISM

New urbanism is as much a revised approach to public policy and governance as a set of principles for making good places. Formed in response to a critique of the problems of the modern city, new urbanism finds the laws that govern urban development largely to blame. Thus while it articulates a set of principles to guide city building, it also advocates a new approach to thinking about the governance of urban form. In this section, I briefly review the principles of new urban approaches developed in the various streams of new urbanism.

Krier's work had an immense impact on Duany and Plater-Zyberk. The classical revival ideas of neo-rationalism clearly inspired the search for traditional principles of design for the building of Seaside. Another influential thread in the new urban approaches came from the interest in sustainable development and healthy communities: movements with many adherents in the mid to late 1980s. This environmentally-based discourse urged lifestyles that demanded fewer resources and that protected opportunities for future generations (WCED 1987). Van der Ryn and Calthorpe wrote:

> these trends will set a new direction for urban design: more compact, mixed-use communities, more efficient buildings, diverse transit systems, an ecologically sound agriculture, water and waste conservation, and ultimately, a greater sensitivity to the uniqueness and integrity of each region.
>
> (Van der Ryn and Calthorpe 1986: ix)

The first substantive effort to set out Duany and Plater-Zyberk's philosophy of neo-traditional town planning came in a book edited by Alex Krieger (1991). As Lennertz (1991:21) wrote, Duany and Plater-Zyberk 'believe design structures functional relationships, quantitatively and qualitatively, and that it is a sophisticated tool whose power exceeds its cosmetic attributes'. Duany and Plater-Zyberk's plans articulate a need for visual and functional coherence. They achieve the objective with a rigorous design code that structures the form of each planned community. Also in 1991, a group that included Duany, Plater-Zyberk, Calthorpe, Peter Katz, Elizabeth Moule, and Stefanos Polyzoides formulated the Ahwahnee principles. The call for complete and integrated communities, mixed use, walkability, and resource conservation as articulated in the Ahwahnee principles (Corbett and Velasquez 2004; Dunlop 1997; Local Government Commission 2002) would soon be echoed in the charter of new urbanism.

New urban approaches commonly focus on the public realm. Good quality design to create attractive streets, parks, and squares is key to the new urbanism. A public realm requires outdoor spaces that generate civic pride and host various uses (Cusumano 2002). It demands walkable streets designed to human scale. Communities need destinations for people to visit and attractive environs along the way. Scully (1994) argues that a focus on urban design is not about taste or preference: indeed, he criticizes Duany and Plater-Zyberk for justifying the use of traditional vernacular simply on the basis that people like it. Good design has intrinsic value that transcends popular taste.

Although new urbanism has drawn some inspiration from theories related to sustainable development, and despite the fact that the adjective 'sustainable' is a popular one for new urbanists to use as a prefix to their solutions (e.g., see Urban Task Force 1999), new urbanism reveals an ambivalence about the place of nature in the city. Calthorpe (1993) suggests, for example, that cities may need to control nature and green space: for instance, he says the garden city offered too much open space. He also argues that designers can put too great a focus on solar orientation: celebrating the urban environment requires towns and town-like spaces. 'Environmentally sound communities need parks, regional greenbelts, and high-quality open space, but they also need density and street-life' (Calthorpe 1993:44). Duany et al. (2000:101) take the argument further: they say that if we are to replace nature with development, then we should do so with pockets of culture (towns, villages, or neighbourhoods). Criticizing landscape architects who too often prettify instead of using vegetation to define urban spaces, the classical revivalists suggest that emulating natural patterns will not prove as visually effective as straight rows of trees in the city.

I use the concept of new urbanism in a broader way than many North Americans might. In much of the writing about new urbanism in North America, the words are capitalized: the New Urbanism. North American New Urbanism is but one variant within a larger set of new urbanisms being practised today. These varying approaches share essential principles while differing in some of their design solutions and terminology.

NEW URBAN APPROACHES

Certain themes unite the new urban approaches. Table 3.1 describes four prominent approaches: traditional neighbourhood design, transit-oriented design, urban villages, and smart growth. Each of these approaches promotes mixed use, mixed housing, compact form, pedestrian orientation, quality urban design, and defined centres and edges for urban neighbourhoods. In the next sections, we review the principles associated with these approaches and note their points of elaboration and difference.

TRADITIONAL NEIGHBOURHOOD DESIGN (TND)

Duany and Plater-Zyberk are best known for having developed the idea of neo-traditional town planning or traditional neighbourhood design. To the classical revival ideas of rational architecture they added a focus on the local vernacular of the regional small town or village. They identified the town centre with its mix of commercial, civic, and residential uses as the heart of community identity and sociability: the centre warrants formal treatment in architecture and street pattern, and a square or green for civic activities. Following Krier (1984a), they proposed putting residential uses over commercial in horizontal layers. They designed a mix of housing types, from apartments over garages to detached houses, with medium densities and low rise throughout. For DPZ, mixing is universally good and can ensure that new neighbourhoods do not contribute to sprawl (Duany *et al.* 2000). Duany and Plater-Zyberk revived the role of alleys and lanes for car access, taking the garage away from the

Table 3.1 Comparing the principles of new urban approaches

Traditional neighbourhood design	Transit-oriented design	Urban villages	Smart growth
Divergent elements of each model:			
focus on vernacular or classical architecture	centred on public transportation hubs linked to regional system	more emphasis on self-sufficiency (with mix of housing and jobs) and brownfield redevelopment	adds government policies and incentives to promote change
Common elements of community design in all models:			
mixed use, mix of housing types, compact form, walkable environment (c. 400 metres centre to edge), transportation alternatives, attractive public realm, quality urban design, centre with commercial and civic uses, clear edges, narrow streets, design charrettes			

front of the house and replacing it with street and lane parking. Most of their designs employ grid or modified grid street layouts intended to give cars and pedestrians route options. Short distances from homes to centre would make it possible for people to choose to walk to key destinations. Front porches on the houses provide transition spaces between the public and private realms and enhance social interaction in the neighbourhoods. Homes push up close to the street edge to give an urban feel. Narrow streets slow traffic. High quality urban design creates a sense of place and history. The TND movement in the USA played an important role in focusing attention on the quality and consistency of urban design quality, even for affordable units. Moreover, the designers insist that there be no visible difference in the quality of housing provided for the poor in mixed developments.

The architectural principles encapsulated in TND presume that people feel comfortable in enclosed spaces with clear signs of human presence (Gindroz 2000). Duany et al. (2000) say that streets need to feel like rooms, and the formal front of houses should receive proper treatment. Visibility, light, openness, order, harmony, human scale, connection, and legibility are the values that inform urban design. Although TND assumes that these values are timeless and universal, we can find abundant cross-cultural and historical evidence to show they are not. Urban traditions in the Middle East and in Asia follow quite different premises (see Chapter 6), and even European cultures change their values through time.

TND developments photograph beautifully. The 'romantic, historicist impression' (Bressi 1994:xxxvi) created through coloured renderings makes the idea of traditional neighbourhoods visually appealing even to those who have never visited Seaside, Kentlands, or Poundbury. The production of coffee table books, like *Seaside* (Mohney and Easterling 1991), played a role in recruiting adherents to the movement.

Although the proponents of TND suggested that it could be used in urban or suburban areas, most of the projects were greenfield developments. This left TND subject to criticism that its shallow urbanism merely facilitated prettier sprawl (Leung 1995). The quality of its design features have made it an expensive option. With its focus on traditional urban form it has become the easiest of the approaches to 'cherry-pick'. Developers have often adopted elements of the design programme – like front porches or back alleys – without buying the entire package, frustrating those who hope to see TND lead to widespread cultural change.

While TND projects are often associated with the USA, neo-traditional developments also appear in Canada, Europe, Asia, and Australia. Traditional patterns and vernacular styles have a powerful appeal in an era when heritage conservation constitutes an important cultural value and place-marketing a significant economic value.

TRANSIT-ORIENTED DESIGN (TOD)

Van der Ryn and Calthorpe (1986), Kelbaugh (1989), and Calthorpe (1993) argued that achieving more sustainable cities would require that designers and urban residents recognize the economic and environmental limits of current development patterns. As a designer, Calthorpe understood the significance of transportation patterns in structuring the urban region. He focused on finding ways to provide viable options to the car so that location and land use pattern would not force people to drive everywhere. The concepts of pedestrian pockets, transit villages, or transit-oriented design looked to public transportation as a development driver.

In some ways, the TOD approach seems more nuanced and less deterministic than the TND concept. It suggests that transportation networks affect land use (and vice versa). To make it possible for people to travel without cars, Calthorpe (1993) proposes to use public transportation systems to structure the region and the neighbourhoods within it. The basic unit of development becomes an area within walking distance of a transportation node (see Figure 3.3). A mix of uses – including offices – occurs close to the transportation station, with density highest near the centre. To the edge of the node are lower density residential uses. The nodes have the 'fundamental qualities of real towns: pedestrian scale, an identifiable center and edge, integrated diversity of use and population, and defined public space' (Calthorpe 1993:33).

The main principles of TOD require a mix of uses and housing types (structured in relationship to distance from the public transportation node), compact form,

Figure 3.3 The transit station as community hub
Trains come into Orenco Station, west of Portland OR. In 2004, the station has some housing nearby, but commercial uses are three blocks away. As the empty fields get developed, the concept of transit-oriented development will turn into reality.

walkable streets, civic and commercial centres along transportation corridors, well-functioning public transportation systems, open space networks, and attractive public spaces. Calthorpe (1993) also favours urban growth boundaries to prevent cities from growing endlessly into the countryside.

Where TND approaches had strong appeal for those keen on classical design, TOD approaches resonated with those concerned about making cities more environmentally responsible. They fused with ideas of healthy cities and sustainable development quite effectively. The presumptions suited models of location choice (Banai 1998), making TOD seem theoretically sophisticated. Implementing TOD did not require allegiance to particular design formulas. Instead it left designers and developers with room to manoeuvre and innovate. These ideas became quite popular through much of the western USA (see Chapter 4) and among larger Canadian cities (see Chapter 7).

URBAN VILLAGES

The urban villages movements adopted many of the principles inherent in TOD. In the American context, urban villages may appear as another name for transit-oriented development nodes that are clearly identifiable, either as infill on a brownfield site or as peripheral new towns along a transit line. Like the other approaches, urban villages promote compact form, mixed use, and mixed housing types. Bohl (2002:109) says: 'When high-density multifamily properties are mixed with retail and other uses, such efforts frequently yield what is characterized as an urban village.' Especially as used in Europe, urban villages offer a strong focus on job creation.

Some may criticize the term as an oxymoron: how can a place be both urban and a village? Yet the ideal of the village, with its defined identity and sociable characteristics, has widespread appeal. Kelbaugh (1997:122) writes, 'urban villages are neighborly. They can create coherent neighborhoods where none exists'. Kelbaugh claims that urban villages are sustainable, and will help to manage environmental problems like air pollution and loss of open space. In the urban village, it appears, one can have the best of both town and country while living responsibly.[2]

In the United Kingdom, interest in urban villages grew out of the Prince of Wales's initiatives. The Urban Villages Forum promoted urban villages as sustainable and relatively self-contained new developments (Franklin and Tait 2002). By the late 1990s, the government had made a commitment to the urban village idea, promising to build Millennium Villages to showcase the most sustainable approaches to urban development (see Figure 3.4). Urban villages would have many features of Krier's urban quarters. They would house 3000 to 5000 people and aim for relatively self-sufficiency (Biddulph 2003a). They would include a focal square or green, be walkable, and enjoy a mix of uses. A primary school, shopping precincts, public transportation networks, and employment centres would give residents the

Figure 3.4 Villages for the Millennium
Greenwich Millennium Village in London is as far from a traditional English village as one could get. Colourful apartment towers and modernist cube townhouses create an urbane ambience in this former industrial area.

opportunity to live near work, school, and shops. Planned open space, environmental conservation, and connected streets with traffic calming would make the neighbourhoods attractive. A high proportion of social housing in the projects (20 per cent or more) would contribute to the stock of affordable housing. While the Urban Villages Forum advocated employing local architectural styles, the Millennium Villages programme held international competitions that were sometimes won by modernist architects.

The urban villages idea has influenced the renewal of town centres and main streets in many communities. It contributes to what in the UK is called the urban renaissance. As a movement, urban revitalization has a long legacy. At one time it involved urban renewal: remove the blight and construct glamorous multi-use centres with malls, convention centres, and hotels. In recent decades, however, many of the principles associated with new urbanism have become intertwined with ideas of town centre redevelopment. The general consensus by the end of the twentieth century suggested that communities should improve their urban centres to generate places for people to meet and be entertained (Bohl 2002; Urban Task Force 1999). The growing commitment to urban revitalization and renaissance focuses attention on stimulating a strong urban mix of employment, commercial, and institutional uses.

THE NEW URBANISM

In the early 1990s, Galen Weston provided funds to facilitate a meeting of those most influential in promoting TND and TOD approaches (Thompson-Fawcett 2003b). These sessions led to the formation of the Congress for the New Urbanism, which in 1993 adopted a charter setting out its principles (CNU 2004a; Leccese and McCormick 2000). The Congress gave the North American movement a unified name, the New Urbanism (always capitalized). It set an agenda for action and research, and worked to reconcile the differences within the TND, TOD, and urban village approaches. As an organizing strategy, it proved extremely effective.

In 1994, Peter Katz produced a book setting out the key ideas of New Urbanism, with chapters by many of the founders of the Congress. North American New Urbanism essentially adopted TND principles at the neighbourhood scale and TOD premises to inform the larger scale (Bressi 1994). Duany and Plater-Zyberk (1994) set out the elements for the neighbourhood, seeing themselves as building on the neighbourhood unit concept (Perry 1929): the neighbourhood has a centre and an edge, an optimal size, a balanced mix of activities, a fine network of interconnecting streets, and identifiable civic spaces. Cities feature multiple neighbourhoods, while a village is a single neighbourhood. Calthorpe (1994) describes the region as including the city, its suburbs, and the natural environment around it. Regions need defined edges, open space networks, and regional transportation systems. In Calthorpe's view, growth boundaries and transportation systems are key to reordering the city. In the New Urbanism, architectural differences between the earlier models were left as questions of 'style' for designers to resolve.

The CNU set out a simple summary of its goals, accompanied by a longer charter that articulates its detailed principles.

> We advocate the restructuring of public policy and development practices to support the following principles: neighborhoods should be diverse in use and population; communities should be designed for the pedestrian and transit as well as the car; cities and towns should be shaped by physically defined and universally accessible public spaces and community institutions; urban places should be framed by architecture and landscape design that celebrate local history, climate, ecology, and building practice.
>
> (CNU 2004a: online)

Annual conferences and regular publications allowed the CNU to establish a growing following, and to persuade designers, planners, and politicians of the worthiness of their cause. The CNU played a significant role in asserting the importance of public over private values in urban design. In making this claim, New Urbanism resurrected the idea of a definable public interest. While the dominant discussion in planning theory in the 1980s and 1990s focused on diversity and plurality, North American New Urbanism increasingly held that the right design could serve the best interests of the entire community.

The components of the good community for New Urbanism were relatively simple. The basic building block, the neighbourhood, links to a centre with shopping, services, and public transportation. The community has a mix of housing and building types constructed according to a set of design principles that reflect timeless wisdom. Designed correctly, and built in sufficient numbers, the product enhances affordability and social diversity, and provides opportunities for social interaction and integration. Transitional areas between buildings and streets (e.g., bay windows, porches, and entrances) help to establish territorial definition and sense of place. Moule and Polyzoides (1994) argued that it takes design, not policy planning, to achieve the objectives of better cities; they add that New Urbanism is not about style but about understanding historical precedents and using them to create a vibrant public realm.

Signing off on the Charter has not fully resolved the differences within the camps of New Urbanism. Calthorpe and Duany still have occasional disagreements. For instance, Calthorpe and Duany recently debated street patterns. Calthorpe proposed one-way couplets to resolve downtown traffic problems in a community, but Duany argued that traffic must travel in both directions for vital city centres (Langdon 2002b). Krier and Duany also differ somewhat on street patterns. We also note differences in policy planning: Calthorpe sees urban growth boundaries as useful, but Duany criticizes them as artificial. The latest debate within the movement involves whether to facilitate access for the mobility-impaired to homes in New Urbanist projects.

North American New Urbanism takes some different positions than we may find in new urbanism applied elsewhere. For instance, Krier has defended meandering street patterns and mews in his urban village at Poundbury as strategies for managing traffic and safety: European new urbanism pays greater attention to facilitating pedestrian connectivity while managing the automobile with courtyard parking. Duany, by contrast, prefers an urbane grid for his projects, insisting on the need for street connectivity.

Calthorpe wrote many books and articles through the 1980s and 1990s to lay the groundwork for his position on urban form in the regional city. In writing about the city, Calthorpe seldom cites contemporary or earlier research to support his positions. Thus he presents a vision of the city rather than a theoretical analysis. The same is true of much of Duany and Plater-Zyberk's and Krier's work. In recent years, Duany has written several pieces to articulate theoretical foundations for the New Urbanism. His articles on the urban-to-rural transect set out a model of urban form that draws on ecological analogies to develop design principles (Duany and Talen 2002a, 2002b). Teaming up with Professor Emily Talen has given Duany's work a theoretical depth not seen in his previous writings. While some of his publications feature assertions, bombast, and hyperbole, Duany's recent articles with Talen employ the kind of rationality that planners understand (Wyatt 2004).

The theory that informs North American New Urbanism is grounded primarily in practical applications. Designers do something noteworthy and then develop theory to explain and substantiate what they have accomplished. The theory is not complex. At times it seems deterministic: good design creates great communities. It offers tempting solutions that respond to the angst we have about the problems of the contemporary city. Its simplicity, and its immediate applicability, give it impact and account for its appeal.

SMART GROWTH

The growing popularity of new urbanism in the late 1990s captured the interest of politicians. As voters in many districts protested against excessive growth that generated urban gridlock, political leaders in North America looked for strategies to make growth more acceptable. The principles of new urbanism proved adaptable to a strategy for good growth: 'Smart Growth'. If growth is properly managed and the correct principles applied, the theory suggests, then growth can help to create better communities.

Smart growth operates from the premise that growth is inevitable, and that if managed correctly it will prove healthy for the economy and the community. The problem with post-war growth, advocates say, is that it was handled badly. Growth management policies that tried to control growth with strict codes and tough zoning bylaws led to urban sprawl, exclusionary policies, long commutes to work, and high housing prices (Downs 1994). Smart growth allows municipalities to plan for growth within the regional urban framework and thus make good use of existing infrastructure while reducing negative externalities. As a theory, smart growth worked well within a culture committed to a free market because it suggested strategies for public policy and intervention that would facilitate development. It caught on quickly in the USA (see Chapter 4).

The principles of smart growth reiterate new urbanism principles, but with less said about design strategies (Smart Growth Network 2004). The extra element added by the smart growth movement involved the idea of using government incentives to encourage the private sector to adopt the principles. The public policy focus of the smart growth movement influenced new urbanism to reinforce the same message.

FROM THEORY TO PRACTICE

By the late twentieth century, then, these various new approaches to urban growth and development appeared to be merging and converging. For planning, the logical next step of adopting the theory of new urbanism and smart growth was to consider adjusting land use policy and regulation to implement the principles. In America, new urbanists argued that inflexible zoning codes presented the greatest barrier to implementing new approaches to urban form. A growing professional consensus brought organizations like the American Planning Association, the Urban Land Institute, the Canadian Institute of Planners, the Prince's Foundation, and the Office of the Deputy Prime Minister of the United Kingdom on board. Governors in several American states, and governments of various stripes in many countries, began to look for ways to implement the new approach. An obvious target was the planning laws and zoning codes set up to deliver garden city principles and values.

Early new urban projects succeeded despite the municipal codes. Their approval often involved months or years of negotiations to provide variances on regulations that would have undermined the innovation necessary for new urbanism to occur. In the late 1990s, though, DPZ began to develop alternative codes, a 'smart code' that would implement the principles of new urbanism over entire districts. Restrictive codes applied to individual properties or districts had enabled projects like Seaside and Poundbury. Municipalities could adopt the new form-based codes to supplement or replace the old 'Euclidean' zoning. The rules were often modelled on historic codes such as the one set for Alexandria, Virginia, that limited building height to the dimensions of the street width (Dover 2004; Lennertz 2000). Codes would articulate a physical vision that land use zones cannot achieve. As Moule and Polyzoides (1994:xxiv) argued: 'The judicious application of codes is to result in a diverse, beautiful and predictable fabric of buildings, open space and landscape that can structure villages, towns, cities and, indeed, the metropolitan region.' While some might argue that such codes would effectively legislate taste (Sitowski 2004), the new urbanists share Scully's (1994:228) view that: 'Without the law there is no peace in the community and no freedom for the individual to live without fear.' Revised codes can balance public and private interests and ensure widespread adoption of the correct design rules to improve communities.

In discussions about the values and principles that should govern the twenty-first century city, we often hear words like vibrancy, vision, density, efficiency, sustainability, livability, walkability, safety, and sociability. Although architects may debate the merits of particular designs, we find widespread agreement about the role that urban design quality and 'visual coherence' play in framing the contemporary city. Every place has to have a 'sense of place'.

How does the theory of new urbanism differ from that of the 'radiant garden city' (as Jane Jacobs liked to call the post-war model)? The garden city planners favoured many of the same principles that animate new urbanism: mixed use, mixed housing, compact form, transportation connections, and civic uses at the community centre. They differed in their approach to density, transportation networks, design quality, arrangement of uses, and development scale. The garden city and neighbourhood unit planners tried to keep cars away from people for safety reasons. Their presumptions about how they might control the automobile proved ill-founded. To remedy the continuing problems, some European cities turned to street-calming strategies in residential areas to improve safety. By contrast, North American new urbanism suggested that improving opportunities for walking, diversifying route choices for cars, and enhancing street design should reduce traffic speed and volume.

The nested hierarchy of relatively self-contained spaces – wherein the smallest is the neighbourhood and the largest is the region – is congruent with the ideas of the early modern town planning movement. Howard's satellite cities formed part of a system of cities linked by public transportation networks. New urbanism has adjusted the spatial pattern and makes different assumptions about household types, but we find many similarities in the theories. In both cases, the theories ignore the broader dimensions of the social and economic order that generate urban problems and patterns. Globalization processes that in the late nineteenth century undermined the living environment of the industrial city continue to affect the fate of the late modern city. Design theories that play with the spatial arrangements of streets, the dimensions of lots, and the heights of buildings ultimately cannot affect the major drivers of urban development and underdevelopment.

New urban approaches have tended to prove more architecturally prescriptive in the USA than elsewhere, perhaps because of the high profile of the TND movement. Urban design has had a boost that it has not enjoyed since the urban nobility of the Renaissance sought to establish its place in the social order through sponsoring building programmes. The grand projects of the early twenty-first century contribute to governments' attempts to improve international urban competitiveness, and to new urban élites' efforts to demonstrate their vision of and role in the good city. To that end, the designers specify building envelopes, massing, setbacks, materials, and the shapes of public spaces.

We do find some differences in new urban approaches around the world, as Chapters 4 through 7 illustrate. For instance, American new urbanism and smart growth tend to employ the grid or modified grid for road layouts. By contrast, the British government seems to prefer winding streets to slow traffic: the mews and cul-de-sacs of the garden city that proved popular in the UK have been integrated into new urban approaches. Also, debates about safety have played an important part in approaches to urban design in the UK. Video surveillance and crime prevention strategies to limit opportunities for crime, have been integrated into public policy. The American approach is rather to promote 'eyes on the street', where perceived surveillance by residents may deter crime. In defence of the American approach, Steuteville (2003b) argues that security is important, but not the only goal in urban design. As is often the case in new urbanism, such debates about strategy proceed with little evidence to indicate which side makes the better factual argument. Case in point: without offering any evidence to support their claim, Duany et al. (2000:73) say: 'In order to discourage crime, a street space must be watched over by buildings with doors and windows facing it. Walls, fences, and padlocks are all less effective at deterring crime than a simple lit window.' Foucault's (1977) premise, that if people believed they are watched then they will behave appropriately, has persuaded the Americans that a window will be enough; the British take even less risk, and employ cameras.

Krieger (1991:13) asks an important question: 'How does one recover the physical planning principles that seem to make good towns without succumbing to mere appearances and producing simulacrum towns?' Marshall (2000) suggests that new urbanism fails in this challenge. To date, he says, new urbanism has focused on style, and ignored policy questions. Marshall believes that governments have to raise gasoline taxes, invest in transit, stop building highways, and control outward growth in order to solve the problems of the city: design alone is not enough.

THE USE OF THEORY

Many new urbanists trained as architects before becoming town planners.[3] As a quasi-profession, planning draws practitioners from eclectic sources. In each case, planners bring the reasoning and theories of their originating disciplines with them, adding to the toolkit they employ. With new urbanism, the literature highlights design rhetoric rather than empirical evidence. As Wyatt (2004) explains, planners and architects think differently. Urban planning draws on the left side of the brain: it seeks to make rational arguments. Planners search for a balanced approach, encouraging full participation by those involved in outcomes, and ensuring careful consideration of all factors before making choices. Architects prefer a different strategy. Architecture draws on the right side of the brain: the creative and introspective approach. Architects may be less concerned with the general public interest than in the needs of particular clients. In the case of the new urbanists, we might argue, the designer-planners use their own creative and introspective processes to determine the public interest perhaps in part to avoid the messiness of participatory evaluation and decision making.

New urbanism often stands accused of making social assumptions and believing that designs can achieve social objectives (Audirac 1999; Audirac and Shermyen 1994; Bookout 1992a; Harvey 1997; McCann 1995). Although the thinking may not be completely deterministic, even cautious authors may end up with phrasing that suggests that spatial patterns create community. For instance, Van der Ryn and Calthorpe (1986:x) write: 'mixed uses draw activities and people together, and shared spaces reestablish community'. Duany et al. (2000:83) say: 'It bears repeating: we shape our cities and then our cities shape us. The choice is ours whether we build subdivisions that debase the human spirit or neighborhoods that nurture sociability and bring out the best in our nature.'

Bohl (2000) claims that the issue is not spatial determinism, but 'environmental affordance'. Design can create spaces that afford opportunities for positive social activities. Good physical environments are a necessary component of the good city, but not the only element required. Bohl believes that critics have used physical determinism as a straw man to discredit new urbanism. While academics like Bohl and Talen deny physical determinism, practitioners seem less cautious. For instance, *Suburban Nation* commonly makes deterministic comments: 'community flourishes best in traditional neighborhoods' (Duany et al. 2000:243). Practitioners often suggest that community can only occur in the right kinds of places: 'poor design – or no design – can discourage human interaction, reduce feelings of community, and provide a sterile, monotonous environment that creates social disassociation and behavioral problems associated with crime and poverty' (Pateman 2004:17). Conventional suburban development, writes Zelenak (2000:online), 'causes not only decay and desertion of the base metropolitan community, but also an increase in crime and the breakdown of basic family values'. Determinism proves difficult to root out of the new urbanism.

Authors like Kunstler (1993, 1999) and Duany et al. (2000) present bold critiques about the way that bad design generates bad behaviour. They argue that suburbs are not good for kids. Children lose their autonomy and become prisoners in safe but unchallenging environments. When kids cannot walk to school, mothers are forced to become chauffeurs. Bored teenagers have no decent public gathering places: hence they get into trouble. They have to go everywhere in cars, and therefore have more accidents. Teen suicides are higher in the suburbs. Growing up in sterile environments makes suburban kids more prone to violence. As Duany et al. (2000:121) say, 'suburban sprawl victimizes America's youth'.

The elderly fare little better. Stranded far from shops, friends, and health care, they have to drive to get anywhere. When they can no longer drive, they may be shipped off to segregated retirement districts, looking for security in gated compounds.

Some of the social assumptions of new urbanism seem based more in hope than in evidence. For instance, the idea that households will accept less land and smaller houses in return for good design or local amenities is not reflected in practice (Audirac 1999). In fact, house sizes are getting larger, even in new urbanist projects in many regions. Will small savings persuade people to give up space? Burchell (2004) says that owners might save $10,000 a unit over 25 years in new urban developments: is that enough to generate a shift in lifestyles? Will saving $6000 a year persuade people to give up the convenience of a car? Responsibility is a hard sell.

We see little evidence that the 'small is beautiful' message will convert people to new urbanism. Indeed, some argue that instead 'luxury fever' has captivated modern society, and is expressed in expensive tastes (Frank 1999). New urbanism may offer an attractive option to those living in a consumer society where housing is a critical social marker: people find homes to announce their place in society. Hence new urbanist projects sell well despite their higher costs (Song and Knaap 2003; Tu and Eppli 1999). A design movement that places shopping facilities at the heart of the neighbourhood, as a 'village centre' for people to meet each other and create community bonds, suits the times. The community becomes the ultimate commodity of the consumer society.

Empirical studies are mixed on the ability of new urbanism to deliver on its objectives. Several studies suggest that new urban form cannot reduce car use as much as its authors may hope (Crane 1996, 1998; Hygeia and REIC 1995; Pogharian 1996). Burchell (2004) says that the new urbanists have failed to provide evidence that parallel parking on streets makes pedestrians safe. Compact development has drawbacks as well as benefits (Breheny 1992; De Roo and Miller 2000). Some researchers have found that older neighbourhoods built according to the principles promoted by new urbanism have higher numbers of pedestrian trips (Greenwald 2003) and a stronger sense of community (Lund 2002, 2003). By contrast, Nasar (2003) discovered that the sense of community was not stronger in traditional developments, although car use was less. The results are thus mixed. Many of the new projects do not have access to shops, transit and other services needed to deliver a level of self-sufficiency (Crane 1996; Grant 2002b; Steiner 1998). Bohl (2000) says that neighbourhoods get the retail they desire: experience suggests otherwise, as lack of suitable stores is a common complaint in developments. Bohl explains the lack of affordability of housing in new urbanist projects as a 'function of the real estate market' (2000:782): the external context makes reaching goals difficult. Both the critics and the proponents of new urbanism admit that some of the basic theoretical premises are not proven, and projects built to date do not always satisfy the movement's principles.

New urbanism advocates providing affordable housing integrated with other uses in neighbourhoods. Units for the poor should be interspersed within market rate housing. The concentration of the poor in public housing projects merits special criticism as one of the greatest failures of post-war modernist planning. In new urbanist discourse, the problems of 'the projects' derive from concentrating large numbers of poor families in modernist high-rise buildings. The physical design created social dysfunction. Duany et al. (2000:53) advocate distributing affordable housing 'among market-rate housing as sparsely as possible in order to avoid neighborhood blight and reinforce positive behavior'. Marxist analyses of poverty might look to the character of the economic system to explain the resilience of poverty in the face of recurrent design interventions in the nineteenth and twentieth centuries, but the new urbanists continue to have faith that good design can resolve the issue.

Thompson-Fawcett (2000:277) says that in the urban village movement in the UK: 'There is an express objective here to encourage people to meet with social difference, in order to enable an understanding of, and empathy for, heterogeneity, and thereby enhance community civility and democracy.' Mixing groups to create social mosaics instead of isolated enclaves promotes urban health in the new urbanism philosophy. As Kelbaugh (1997:123) says:

> Although this mixing may be less comfortable and genteel than gated subdivisions, social insulation slowly builds the kinds of alienation that ultimately erupts in civil violence ... it is better to take our legitimate differences and frustrations out on each other in small, everyday doses than in episodic racial and class upheavals. The more personal interaction certainly offers greater hope of understanding among human beings.

In the absence of empirical analyses that demonstrates that mixing can overcome societal differences and inequities, we are expected to have faith that the minor inconveniences are worth the trouble, or indeed that they may prevent incipient insurrections. In practice, of course, as Talen (1999) notes, most new urbanist projects have proven relatively homogeneous in composition. Rather than minimizing difference, new urbanist projects have sometimes exacerbated it by creating enclaves of affluence in the urban environment.

While new urbanism lauds diversity, we might argue that in some ways its approach to difference is assimilationist. That is, the civilizing project of the redesigned community involves finding ways to incorporate those whose behaviour may otherwise threaten or disrupt urban harmony. Through propinquity, the poor interact with and gradually come to emulate the households around them; thus they learn to minimize their difference. Instead of accepting difference, we transform it.

Mixing income levels contributes to a debate within new urbanism. If housing values begin to rise, and poorer households find it difficult to obtain housing in the neighbourhood, is that good or bad for the community? Duany et al. (2000) suggest that neighbourhoods can accommodate an abundant mix of uses and building types, but not of people: a maximum of 10 per cent affordable units provides a good rule of thumb for developers. For Duany, gentrification constitutes a natural part of the evolution of cities: it happens to well-designed urban areas. 'It is the rising tide that lifts all ships ... What spokesmen for the poor insist on calling gentrification is actually the timeless urban cycle of a free society organically adjusting its habitat' (Duany 2000:online). This theory assumes and values social mobility: the design process facilitates it. Interventions by government to prevent the market from operating are doomed to fail. Increasing property values in a neighbourhood are seen as 'stabilizing', as a developer said about a Portland, Oregon, project: 'Our first phase helped greatly to stabilize the neighborhood, ... prices for homes in the neighborhood have almost doubled since Phase I was completed two years ago' (Lassar 2001:46). As we will see in the discussion of the renewal of public housing projects in the USA known as HOPE VI (in Chapter 4), the need for mix has provided a key rationale for destroying public housing units even at a time when tens of thousands of Americans are homeless. A sceptic might conclude that some proponents of new urbanism seem influenced by libertarian values.

While new urbanism values diversity, it also has made developing and strengthening community a key goal. Talen (1999:1361) admits that new urbanism 'lives by an unswerving belief in the ability of the built environment to create a "sense of community"'. Its claims are hardly modest. She notes that new urbanists hope to build a sense of community by good design and placement of public space. Designers commonly talk about features such as porches, gathering places, wide sidewalks, and mixed use as contributing to the development of community. As Talen says, some of these claims prove hard to substantiate. She urges the movement to avoid the language of 'sense of community'. Since, as she notes, homogeneous communities often have a stronger sense of community than do heterogeneous ones (even though they fail to meet the tests that new urbanism sees as key to good communities), perhaps new urbanists should focus on sense of place or a notion of the common good instead of sense of community. 'New Urbanists have translated the building of sense of community into a specific design manifesto' (Talen 2000a:173). At the core of the issue, she finds three social goals: social diversity, accessibility, and local identity (Talen 2001).

> The use of the notion of community has opened the door wide to critics of otherwise laudable town-planning theories (such as traditional neighborhood design), critics who are quick to point out that physical determinism has never been morally or practically supportable.
>
> (Talen 2000a:179)

Talen warns new urbanist designers and developers: 'Focusing on slippery issues like community, as customarily defined, plays into the hands of critics bent on finding reasons to mistrust New Urbanism' (2001:online). She suggests that developers talk about the 'common good' instead of focusing on social relationships as the outcome of new urbanism. She believes that appropriate design represents a common good that supports community by providing places where people can interact with others. Once again, we see that despite an espoused commitment to diversity, the new urbanism retains at its core a faith in a unitary public interest.

Although new urbanism is clearly a design movement at its heart, the question of *style* remains a difficult one for the movement to resolve. Krier (1984a, 1998) is most explicit in arguing that mixing styles produces chaos and discordant environments; he prefers continuity in style to generate order and beauty. Duany *et al.* (2000) see this as especially important for providing housing for a mix of income levels. Some proponents of new urbanism prove sensitive to accusations of nostalgia (Ellis 2002). They dismiss critiques that adherence to classical or traditional principles reflects a lack of creativity. Increasingly the material produced by the Congress for the New Urbanism argues that style is not the issue: that urban design is interested in the shaping of spaces, not the details of the buildings that shape space. Yet the codes that provide essential implementation tools for new urbanist projects typically deal with style. They shape the building envelope and in many cases encourage builders to select local vernacular materials, colours, and styles.

Duany *et al.* (2000) favour architectural codes to develop harmony of style. They say that: 'Traditional neighborhood design has little or nothing to do with the issue of architectural style' (2000:208). This suggests that they separate the traditional patterns of town layout from the design characteristics of particular buildings. As they acknowledge, though, the market responds most effectively to traditional style in buildings. The architectural establishment dismisses classical architecture for the wrong reasons, say the new urbanists. DPZ and Krier see classical style as a suitable tool to use to market their bigger design ideas of density, connectivity, and an attractive public realm. New urbanists view design as one of the requirements for creating good communities: success also takes investment (economic and social), participation and vision, and regional planning (Calthorpe and Fulton 2001; Duany *et al.* 2000; Ewing 1996; Fulton 1996).

The new urbanist design process is participatory in that it involves stakeholders and interested community members in the design charrette that sets out the vision for the community. As Talen (2002:183) says, 'participatory processes in New Urbanism, defined in terms of enhancing social interaction and building consensus, form a significant part of the community-building efforts of New Urbanism'. We note, however, that this level of involvement in a preliminary phase of the planning process is certainly not the kind of participatory democracy that collaborative planning theorists advocate. Once the design is finished and the codes set, where are the opportunities for further democratic action? New urbanism presumes that good design obviates the need for further citizen participation. Duany *et al.* (2000) argue that happy citizens don't protest. Good design will make people happy. Community efforts to resist projects or to change policy are generally dismissed as NIMBYism, or as evidence of a need for better public education about the requirements of good communities.

In contemporary culture, we have great faith in education as the cure for most ills. New urbanism falls victim to the same hope. With education, people will come to understand that new urbanism is the best chance for our cities (see, e.g., Talen 2001). As Van Tilburg (2000: 58) writes: 'Professionals in the architectural field have the opportunity to confront public misconceptions about higher density with a powerful tool – design. Images can make a compelling statement that can overcome even the toughest mental block.' Differences of opinions thus reflect either ignorance or stubbornness, not reasoned choice. Designers often argue that people simply need to see examples of the new urban styles to realize that the alternative improves on the conventional. They assert that people choose the house with the garage in front because too few houses have garages in the lanes. Of course, the reality is seldom so simple.

New urbanism offers a *normative vision* that Talen (2001) believes planning requires. Planning is not just about managing traffic, issuing development permits, and holding hearings on applications. Planning involves thinking about the kind of future we face, and the kinds of places we want to live in. Talen argues that planners need to give people a positive vision and the tools to achieve it. She suggests that the components of a theory of good urban form would draw on the work of Jane Jacobs, Kevin Lynch, Lewis Mumford (only for the regional planning piece), Leon Krier (the only non-American in her list), and Christopher Alexander. Talen and Ellis (2002:41) acknowledge that authors may have divergent images of the good city of the future: 'A central, daunting question has been, whose normative vision is to be adopted? . . . on the plane of fundamental spatial principles, some theories are just better than others.' For them, the answer to 'whose vision' is simple: designers trained in an understanding of good urban form should identify the better principles for citizens. Although the visioning process may involve local residents, the normative vision to guide development practices involves a professional perspective. I will return to this theme of where power lies in the new urbanist model after considering the place of ecological theory in new urbanism.

THE USE OF ECOLOGY

Since the nineteenth century, evolutionary theory has held a place of high repute in the sciences and social sciences, often employed in completely inappropriate ways (Sahlins 1976). In contemporary times, we see evidence that ecological theory has a comparable fascination for authors, and is similarly abused. New urbanism commonly uses ecological language and organic analogies, and in recent years has sought to adapt ecological theory to use in determining urban form.

Thompson-Fawcett (1998) has explored Krier's use of the organic analogy. For Krier, the city as organism represents a significant transformation in thinking from the modernist world view that conceived of the city as a machine for living. 'Against the amorphous social and physical form of functional zones and neighbourhoods, the quarter represents a social organism of definite physical size' (Krier 1978:163). The city should grow through a process of clonal replication, not mechanical reproduction (Krier 1998). Similar analogies infuse the work of Talen, who says, for instance, that new urbanism focuses on 'organic patterns and interrelationships' (2000a:179). Although for the new urbanists, architecture represents the triumph of culture over nature, organic analogies feature rhetorically as a way to naturalize prescriptions.

Explicit references to ecological theory and thinking are used selectively to drive theoretical development in new urbanism. For instance, Calthorpe (1993:11) says that ecology provides the real counterpoint to modernism: it challenges traditional ways of doing things and forces us to reconsider our choices. A few pages later, though, Calthorpe accepts that design involves mapping out a 'new direction for growth' (1993:15). Ecologists might argue that limitless growth contravenes the principles that drive ecological systems (Wackernagel and Rees 1995), but new urbanists seek to accommodate growth. What does a new urbanist mean when he writes: 'Nature should provide the order and underlying structure of the metropolis' (Calthorpe 1993:25)? Ecological models provide a way for new urbanists to think about the interconnections that affect the livability and economic success of regions, but they are seldom used to challenge the fundamental premises of the development industry.[4] As Bressi (1994) has acknowledged, new urbanism has often failed to respond adequately to ecological concerns. For instance, new urban designs for compact development may leave huge swaths of landscape bereft of ecological function. The October 2004 conference on 'A Vision of Europe', in Bologna, had as its theme, 'New civic architecture: the ecological alternative to sub-urbanization' (AVOE 2004:online). New urbanism applies ecology in a strictly limited way, often as legitimizing theory, stripped of challenges to the designer's vision.

The rural to urban transect, formulated by DPZ, is presented as a theoretical underpinning of the new urbanist approach. As early as 1994(xvii), Duany and Plater-Zyberk wrote: 'Like the habitat of any species, the neighborhood possesses a natural logic that can be described in physical terms.' Rather than treat the city as an artifact of human choice, they suggest an essential or natural logic to the built form. Drawing explicitly on the analogy of the ecosystem, the transect looks at the city as a set of zones or habitats with different structures and composition. 'Based on ecological theory, the transect is a regulatory code that promotes an urban pattern that is sustainable, coherent in design, and composed of an array of livable, humane environments satisfying a range of human needs' (Duany and Talen 2002a:245). New urbanists apply the transect, translated into the copyrighted SmartCode, to modify land use regulations in several North American cities. The theory seeks less to explain the urban environment than to provide the rationale for prescriptions about mixed use and design types. The language of ecology serves a strategy for articulating a normative vision of the city; a sceptic might suggest that its primary purpose is as a marketing tool for the designer's preference.

Talen has worked assiduously to develop theory for new urbanism. In one paper, she wrote, 'new urbanism may be with a subcategory of human ecology known as "environmental sociology", which has its roots in the theoretical model of Talcott Parsons' (Talen 2000a:173). As Duerkson (2004) notes, one can also see evidence of the Chicago School's concentric zone theory of the city in the transect model (see Figure 3.5). More recently, in collaboration with Ellis, Talen found inspiration in Frederick Turner's (1991) theories about intrinsic human nature and shared concepts of beauty and culture. 'Some (Turner in particular) have pursued the theme that evolutionary theory, chaos theory, and classicism can be tied together to spur a renewed attention to durable, time-tested cultural patterns grounded in our evolutionary history and neurobiological structure' (Talen and Ellis 2002:40). Talen and Ellis appear ready to apply sociobiology in planning theory. A popular theory in pop culture in the 1960s and 1970s, sociobiology held that human biology accounts for a significant proportion of human behaviour. Desmond Morris (1968, 1969) made a fortune from books that explained the loutish behaviour of beer-sotted brutes ('man the hunter') as the environmental legacy of millions of years of a hunting existence. Such essentialist reasoning assumes that particular cultural values and behaviours are 'intrinsic' or part of our 'human nature'. The values postulated as normal, natural, or universal, coincidentally, tend to be those held by the authors. Any examination of

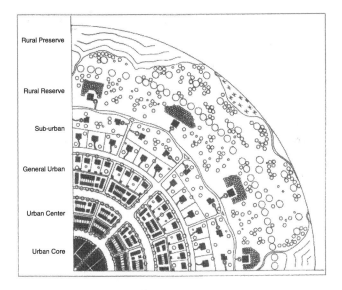

Figure 3.5 Transect/concentric zone theory
In some ways, the transect model maps effectively onto the earlier concentric zone model of urban growth. While the earlier theory focused on use – with the CBD in the centre and commuter zone as the fifth ring – the new one concentrates on form. In both cases, the pattern correlates to distance from the urban centre.

the extent of cultural variation around such values and behaviours immediately dispels any myth that humans share essential characteristics. Variation is the human condition. In the contemporary era, we resist explanations that might claim that gender determines capabilities, or race fixes intelligence, yet we are expected to accept that our essential nature makes us prefer classical columns on buildings. Such thinking defies logic. Cultural myopia masquerades as universal values.

THE ISSUE OF POWER

New urbanism has tended to overlook the possibility that the city takes the shape it does principally because that form has suited real estate interests eager to increase returns on investment (Foglesong 1986; Harvey 1994, 1997). Fainstein (2000) suggests that new urbanism has little concern for social justice and ignores issues of power; its benefits are exaggerated and limited to a small class of urban residents and developers; its participatory methods coopt participants and manipulate consensus-building to suit the interests of a small segment of society. Marxist analyses suggest that new urbanism serves primarily as a way of marketing urban development to promote a particular ideology.

A strong authoritarian streak permeates new urbanism. Powerful investors, including a prince, have used new urban approaches to disseminate their personal prescriptions for urban life. Even more influential, though, are the designers (Teitz 1998). Krier (1984a:38–9) wrote, with a flourish and bold capitals, of the need to 'establish the Authority of Architecture as Art'. The architect has the expertise and vision to set the urban agenda. With new urbanism, 'architecture regains its traditional stature as the means by which cities are made' (Scully 1994:225). The era of patronage when European princes and popes commissioned architects to create culture enjoys a reprise. As Dumreicher *et al.* (2000:293) say: 'In spite of the fact that the sustainability design process will require working in a highly interactive way with other professionals and stakeholders, the architect will be located in a more critical position than conventional practice affords.' New urbanism (re)positions architects/designers in a central place in the process of shaping cities and their culture.

Even Bohl (2000) acknowledges, however, the risk that designers may manipulate the process. They are educating as they are engaging the public, explaining the nature of 'good design' and elegant proportions. They have specialized knowledge and insight: expertise provides power. In the charrettes, they hold the drawing tools. They formulate the codes. Their élite values inform the process from start to finish, yet their power remains largely unacknowledged.

If classical principles constitute universal elements of beauty – as new urbanism suggests – and if humans by nature crave beauty, then the experts who understand the rules of beauty gain legitimacy to prescribe design solutions. In privileging the visual over the verbal (Moule and Polyzoides 1994), new urbanism also elevates the designer over the manager or planner. While collaborative planning theory describes the planner as a facilitator or enabler, new urbanist approaches have planners/designers as experts endorsing solutions. 'Universal truths' and notions of essential 'human nature' serve the interests of power: they deny diversity and trivialize dissent as ignorance (see Horne 1994). The balance of power shifts dramatically from the people to the expert. In this, new urbanism operates in an analogous way to rational planning.

Some critics of new urbanism accuse its proponents of class bias. For instance, George Will (2004:64) dismisses those who wage war on Wal-Mart; he decries the contempt of progressives for 'the vulgarity of popular tastes'. Baxandall and Ewen (2000) describe a profound anti-suburban bias in our culture: a snobbery or class-based antipathy. Their research describes the suburbs as places that provide decent housing, schools, and parks to the masses. Moreover, they see the suburbs as diverse and caring places: real communities.

Some may argue that new urbanism cloaks issues of power in persuasive rhetoric. Marshall (2000) says that new urbanism corrupts language by building 'town halls' to house management offices and residents' associations, and by calling suburbs 'villages'. Residents' associations may receive democratic sounding names, as in 'The Kentlands Citizens Assembly'. Thus the new urbanism coopts the language of New England self-governing communities as it generates post-modernist private management (Marshall 2000; McKenzie 1994). It employs the terminology of consensus-building as a means to its designers' preferred ends. It adopts the rhetoric of sustainability (itself a poorly articulated set of theories) even as it promotes an agenda of growth. Its intellectual honesty may be suspect.

Do new urban approaches constitute a break from modernism? On the one hand, they certainly represent a powerful critique of modernist principles and a resurgence of pre-modernist premises. On the other hand, new urbanism does not completely abandon modernist values or approaches (Beauregard 2002; Ford 2001). It retains a commitment to the model of planner/designer as expert. It relies on an expanding economy. It continues to try to accommodate the car and rising consumer expectations. It employs codes and rules to order society. Banai (1996:184) believes that 'the neotraditional town can be seen as a (piecemeal) spatial strategy to create some stability by means of historicized symbolic landmarks to fix the suburbs'. It represents a reform of the modern settlement pattern rather than a replacement of it. We might call it a 'traditionalized modernism' rather than a return to traditional values.

New urbanism is more than just an approach that draws on historical precedents to create better neighbourhoods as Bohl (2000) would have us believe. It comprises an effort to promote the good community through setting out the spatial conditions for generating desired social behaviours and connections. After recognizing that the modernist city failed to deliver the quality of space possible and the quality of life desired, new urbanism has postulated alternative principles that in times past generated livable cities. Some of the objectives new urbanism promotes have been met in practice; others have not. The debate about the utility of new urban approaches continues, even as more projects are built each year.

Despite the criticisms, new urbanism has had a significant impact in the market. It influences development in a wide array of cities and countries. In the next section, we will look at examples to see how new urbanism is interpreted and implemented in practice. How does it translate into a range of cultural contexts and histories? Can an idea that draws on the European urban experience, and responds to the American urban context, make its way around the world in the way that the garden city model did? While the review that follows is far from comprehensive, we will consider the way that new urbanism is informing the discussion of how we might achieve the good community. From practice we look for insights to inform theory.

PART 2

NEW URBANISM(S) IN PRACTICE

CHAPTER 4

NEW URBANISM IS BORN: THE AMERICAN EXPERIENCE

Transforming a history of sprawl

Every year since the *New Urban News* began publishing, the number of new urbanist projects in the United States has increased. By 2004, projects were up by 37 per cent over the year previous (Steuteville 2004a:1). Steuteville reported 648 neighbourhood-scale communities in the works, with 369 actually built or under construction. A neighbourhood scale project includes at least six hectares with a central gathering place and the potential walk to shops or businesses when complete. The largest new urbanist projects, like Celebration in Florida, expect to house about 10,000 people.

Some developers see new urbanism as the 'Next Big Thing,' ready to take over the development industry (Contreras and Villano 2002:36). New urbanist projects have certainly received abundant media attention. An idea that began with a plan for a coastal resort in Florida has blossomed into a full-blown movement with theorists, designers, and builders making their mark on American communities. The Congress for the New Urbanism, formed in the early 1990s, has become a powerful voice for change in American cities. Professional organizations of varying stripes have found new urbanism and smart growth alluring, and have promoted the ideas extensively to their members. Even a segment of the political infrastructure of American society adopted the principles of new urbanism to support the movement for smart growth. In this chapter, I discuss some of the new urbanist communities in the United States, and consider the challenges that new urbanism can and cannot address.

Rapid urban growth in the USA in the post-war period led to a general consensus by the 1970s that American cities were sprawling out of control. Sprawl set the context in which new urbanism developed as a compelling alternative to business as usual. The basic principles of new urbanism were becoming clear by the early 1980s. Grounded in the ideas of Jane Jacobs, Christopher Alexander, and Kevin Lynch, the principles of new urbanism suited a period when, with Ronald Reagan as President, America found new appeal in traditional values.

When Leon Krier came to the USA to lecture architects about classical architectural principles in the late 1970s, he found a receptive audience in the Florida-based architects, Andres Duany and Elizabeth Plater-Zyberk. They, in turn, taught a generation of students at the University of Miami – architects like Victor Dover, Joe Kohl, Jaime Correa, and William Lennertz – who went on to spread the word (Dunlop

1997). On the other coast, Peter Calthorpe and his colleagues were testing ideas about how to promote urban sustainability within a regional context. These individuals became key players in giving form to new urbanist ideas in practice in America.

The ideas that inspire new urbanism are not new. Houstoun (2004) notes that an advisory panel to the Urban Land Institute (ULI) and the Housing and Urban Development department (HUD) in 1981 recommended ideas now at the heart of smart growth. Concern about sprawl dates back at least a century. American new urbanism built on a deeply-rooted tradition of thinking within architecture and planning. The movement has continued to develop and to expand its influence over the last two-and-a-half decades. It seems safe to assert that in the United States new urbanism has become a national design movement while smart growth has become a national political movement.

In a single chapter, I cannot review all the new urbanist projects in America. That history is waiting to be written. Instead, I will highlight some of the more interesting and famous projects, and draw some lessons from the American experience. The examples illustrate how great an impact new urban approaches can have, but also the difficulty of transforming contemporary development patterns.

AMERICAN NEW URBANISM PROJECTS

The two most famous new urbanism projects – Seaside and Celebration – are located in Florida, where the movement has had a considerable impact on built form. The original neo-traditional village was Seaside, on the Gulf Coast. Developer Robert Davis hired Andres Duany and Elizabeth Plater-Zyberk, then in the process of launching their own architecture and planning firm, DPZ, to do the plan. Davis and his designers wanted to create a real town, with a mix of uses and people, instead of a mere cluster of cottages. In 1982, Seaside began to take shape, and in the process helped to spawn a movement.

The Seaside Institute (2004) suggests that at about 32 ha (80 acres), Seaside is Krier's ideal size for an urban quarter. The designers hoped to replicate the authentic urbanism of southern towns: they looked to those communities for inspiration for the design codes that would govern building. Mohney (1991:36) argues that 'Seaside was a serious attempt to address the issue of the public realm', to create a beautiful as well as functional community. Mohney and Easterling's (1991) well-illustrated book demonstrates how picturesque the place proved. Seaside even served as the set location for the movie, *The Truman Show*. If Celebration is the most famous new urbanist town because of its Disney connections, Seaside is the most photographed because of its simple yet classic beauty.

The street layout of Seaside reflects Beaux Arts influences, with geometric patterns at key civic locations and diagonals to facilitate access (see Figure 4.1). Instead of the huge lots that had come to dominate American suburbia, Seaside offered small lots, mid-block walkways, narrow but connected streets, shared access to beach, and village centre with shops. To achieve harmony in architectural form, Duany and Plater-Zyberk formulated a design code that specified the relationship of building to street and other features key to establishing a sense of place (Krieger 1991). They left the design of individual buildings to others to encourage variety.

Seaside proved financially successful beyond its designers' dreams. The community experienced rapid and profound inflation in property values. While lots originally sold for moderate prices, with the initial hope that apartments over stores and over garages might provide affordable housing, instead the village became a seasonal resort for the extremely affluent. Even small cottages sell for more than $600,000 or rent for $1200 a week. Although Mohney (1991:46) argued that 'Seaside strives to avoid pernicious elitism', it has not escaped its own success.

Other new urbanist projects have developed near Seaside. WaterColor, a 202 ha community designed by Jaque Robertson, is growing next door through the efforts of the St Joe Company (30-A.com 2004). Aurbach (2000) reports that Seaside residents resisted street connections to this new development. Although the idea of street connectivity pervades new urbanism principles, like the occupants of gated

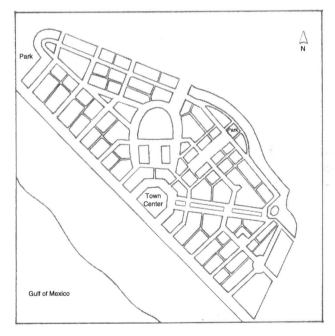

Figure 4.1 Seaside site plan
The layout of Seaside reflects a simple grid enhanced with Beaux Arts elements at its civic core.

enclaves the residents of new urbanism communities may seek to limit traffic on their streets. Practice challenges the principles of the paradigm.

Having seen the potential of neo-traditional town planning at Seaside, in 1989 Canadian billionaire Galen Weston,[1] decided to build a private golf community on a barrier island near Vero Beach, Florida. Weston hired DPZ to prepare the plan and design codes for Windsor (Windsor web site 2004). Windsor enjoys a mix of uses around the golf course: air strip, gun club, store, inn, and post office. Homes range from $1 million to $8 million. The design attempts to create a village setting where people can walk about and interact with each other. Lots are fairly small for that level of affluence. As Strupat (2001:online) suggests, though, Windsor is 'lived in for only a few months each year by like-minded rich people': a pleasant get-away from the demands of life. It is hardly a balanced or complete community.

Strupat (2001) claims that Windsor has gates to limit access. If true, then Windsor reveals a dilemma that new urbanism faces in practice. The affluent in America find solace in separating themselves from others. Maintaining street connectivity and attaining social and economic mix is a significant challenge given the nature of project financing and consumer preferences. While the neo-traditional elements of new urbanism prove immensely popular with certain purchasers, some of the social and spatial objectives of the movement prove difficult to implement.

Peter Calthorpe designed a new community at Laguna West, near Sacramento (California), that also contributed to the discussion about new urban approaches. Construction began on the 324 ha site in 1990 (Staley 1997). The plan called for 2300 units around lakes, joined to a commercial centre in a traditional town design (Calthorpe 1993). Good connectivity and design would create a walkable place. Calthorpe hoped Laguna West would eventually link to the city by rapid transit. Apple Computers located there, giving the community an important employment base (Calthorpe 1993).

To meet market demands in the area, Calthorpe had to redesign the project from his original concept. The final plan resulted in more cul-de-sacs than usual in a new urbanist community. The project sold more slowly than expected, leading the developer into financial trouble. Laguna West has somewhat higher densities than conventional development, but otherwise is similar to other California suburbs. Some homes have front porches, but garage-front homes also appear (Cox 2002). Like Seaside, the project is walkable, but people live car-oriented lives. The community is as often described as a failure as it is called a success.

Located in commuting distance of both Baltimore (Maryland) and Washington DC, Kentlands[2] in Gaithersburg (Maryland) provides one of the clearer success stories of new urbanism. A year-round community with a vibrant mix of uses and housing types, Kentlands demonstrates that with good design, higher densities create attractive and livable settings (see Box 4.1).

Box 4.1 Kentlands: American dreams and realities

In the late 1980s, developer Joe Alfandre approached urban designer, Andres Duany, and asked what might be done with a large tract of land outside Washington DC. Alfandre had previously received proposals from design firms that fell short of his expectations. Although he did not yet have a clear idea of what to build, cautious deviations from the conventional norm would not suffice. With his recent work in Seaside and evolving thoughts on new urbanism, Duany provided Alfandre the assistance he sought. By the early 1990s, the erstwhile farm known as Kentlands had become the leading model for advocates of new urbanism.

When working as a municipal planner in Maryland, I spent an afternoon with Alfandre looking at potential sites for redevelopment outside Baltimore. A creative optimist by nature, Alfandre can, nevertheless, offer a stinging critique of developers, elements of which are part of the larger story explaining why relatively few projects like Kentlands have broken ground over the past decade. He laments the fact that Kentlands remains such a prominent model of new urbanism. In his mind, that torch should have been passed long ago.

Today, Alfandre's 'model' is a community filled largely with highwage earners who work in a Washington metro area bogged down by traffic and saddled with high housing costs. The prevalence of Mercedes Benz automobiles in Kentlands testifies to its appeal, as does the rate at which houses appreciate: consistently higher than comparable conventional developments in the surrounding area.

Kentlands' designers and builders did in fact do most things well. Public space is creatively woven through neighbourhoods, civic buildings terminate vistas, and people have an opportunity to walk through nature parks or to shopping districts without feeling out of place. Unfortunately, few if any who live in Kentlands have an opportunity to walk to work, since the kind of office space high wage earners typically occupy does not exist. One must drive to neighbouring King Farm to see an example of office buildings integrated into a new urbanist community.

The main street is economically viable, if not particularly vibrant. Higher density housing is appropriately located above shops and adjacent to the main street commercial district. The occasionally maligned conventional commercial shopping centre is actually reasonably well-integrated into the community in terms of walkability and scale. Infill development is taking place along the edges of the shopping centre, creating a main street district along a road that had previously only seen the backs of K-Mart and the local supermarket.

Despite Kentlands' many successes, I have never left the place feeling elated. Part of my discomfort runs parallel to Alfandre's. The better part of two decades have passed since Alfandre and a cadre of like-minded urban designers used design charrettes as a tool to work with the public and elected officials to hammer out the details of a plan based on traditional planning principles. The level of interest and energy required to drive these efforts can exceed what is available on planning boards, elected councils, and their appointed planning directors. From a political perspective, it is often easier to have staff generate zoning ordinances and maps to accommodate growth.

> My other source of discomfort stems from issues that go beyond urban form. In the United States, local governments with the means embrace what Robert Kaplan (1998) calls 'jurisdictional sovereignty' to separate the affluent from problems typically associated with the African-American majority in cities and economically distressed areas. Gaithersburg, the jurisdiction in which Kentlands sits, illustrates his point. Policy instruments, such as community land trusts, used to preserve housing affordability, are non-existent. As such, many of the socially-oriented goals formally embraced by the new urbanism are orphans of an increasingly polarized and fractured society. Kentlands is a community only for those who thrive in the modern economy.
>
> As a resident of a transitional neighbourhood in Washington, I find that the depths of problems 'jurisdictional sovereignty' seeks to shut out are challenging for me to ignore, especially in the context of raising children. The despair, violence, and dysfunction commonplace not only in Washington, but also in other impoverished communities across America, are extreme by western standards. Fear is a palpable by-product of historical events and circumstances. When faced with a hypothetical choice between Kentlands' exclusivity and the disturbing realities of urban America, I find neither appealing.
>
> Source: Patrick Moan

On the 143 ha site, Duany and Plater-Zyberk designed Kentlands to include a large commercial area (regional mall), and a mix of housing types from single detached estate homes to apartments over garages. Kentlands has a church, school, community centre, and lots of green space. Main Street shops in a traditional style screen the mall from the residential areas of the town, but seem a long distance from some of the neighbourhoods.

With Patrick Moan and some colleagues from a workshop at the National Center for Smart Growth Education and Research at the University of Maryland, I visited Kentlands on Saturday, 2 October 2004. Although it was a lovely autumn day, we found little activity on the traditional-style Main Street (see Figure 4.2). Cars passed by regularly, and many vehicles parked along the street, but few pedestrians walked about. By contrast, the retail malls on either side of Kentlands Boulevard bustled with weekend shoppers arriving by car. Main Street features mostly service uses: real estate agents, doctors' offices, restaurants, and the like.[3] The malls, on the other hand, had grocery stores, big-box retailers, and fast food outlets. Consumers seemed to favour the malls.

Kentlands is a beautiful community, with lovely small parks and well-kept properties (see Figure 4.3). The vegetation is maturing nicely, giving the place a homey feel. As we strolled the neighbourhoods, we met few residents out walking, despite the glorious weather. One woman, from whom we asked directions, told us how much she loved the community: she felt comfortable letting her children ride around

Figure 4.2 Kentlands Main Street
On a pleasant October Saturday afternoon, Kentlands Main Street proved very quiet. Its shops with apartments above create a lovely ambience, but do not entice many pedestrians.

Figure 4.3 Kentlands townhouses
Homes in Kentlands are universally attractive and well-kept (here on a street named after the developer). Landscaping standards are high.

on their bicycles, and she had great relations with others in the neighbourhood. At Hallowe'en, she said, Kentlands proves a magical place because so many people decorate their homes and yards for the children.

Kim and Kaplan (2004) found a higher level of attachment and community identity in Kentlands than in a neighbouring conventional development: the areas had comparable house prices, but different mixes and street patterns. Prices in Kentlands, and in the Gaithersburg area generally, are not cheap: it proves hard to find a townhouse for less than $500,000. That reduces the amount of economic and ethnic diversity found there. Also, the compact form with small setbacks from street and neighbours, creates challenges for privacy. Nevertheless, the project has sold very well.

Kentlands offers one of the best examples of using new urbanist ideas to create more attractive suburbs. While Kentlands aspires to be a town, it is not. Although its designers hoped building a Main Street would create a vibrant town centre, and generating a pedestrian-friendly streetscape would encourage people to walk for their shopping needs, the reality we saw was people driving to the big box centre. Children and seniors are likely the main beneficiaries of the spatial patterns created at Kentlands: kids can bike or walk to school, and seniors' apartments abut the well-designed shopping districts. Design can create some advantages in a place like Kentlands, but alone it cannot shape travel patterns or social mix.

With the notoriety gained by early projects like Seaside, Laguna West, and Kentlands, some of the key designers and financiers of neo-traditional town planning and advocates of transit-oriented development met to consider how to advance the movement. The initial meetings of the Congress for the New Urbanism (CNU) were limited to the invited. Kelbaugh (1997:133) says this was not élitism: but being too open 'ends up in pluralist confusion'. The CNU generated a unified term for the movement, 'The New Urbanism', and provided a consolidated context for spreading the word throughout the USA.

About the time that the CNU was organizing, Celebration took shape. The Disney Corporation had extra land around its theme parks in Osceola County, Florida, and decided to build a new town. The project broke ground in 1994 (Ross 1999). With Disney resources behind it, Celebration got the full new urbanist treatment, although the planners didn't use the term. Designers Robert Stern and Jaquelin Robertson set out a lazy grid around a lake, with a town centre of mixed uses. While DPZ projects rely on codes to control form, Celebration ensures conformity through a pattern book for homes. Builders must select from a limited template of home plans. Visits to southern towns and reviews of *Southern Living* magazine inspired the designs (Frantz and Collins 1999).

'Celebration is a picture-book town with a Kodak moment on every corner, and it reminds all first-time visitors of a film set' (Ross 1999:19–20). The town proved successful for Disney, attracting many Disneyphiles, as well as families looking for an alternative to conventional suburbs. As Marshall (2000:xviii) notes, though: 'Celebration is a conventional suburban subdivision pretending to be a small town.' It lacks the diversity and messiness of a real town (Frantz and Collins 1999). It has no affordable housing. The retail uses in the town centre built by Disney have been supported in part by tourists, but do not meet the daily needs of residents. Frantz and Collins (1999) complained, for instance, about the lack of a bookstore. Every day shopping requires residents to get in their cars to drive to the mall.

By the later 1990s, new urbanist projects appeared across the country. King Farm in Rockville, Maryland, started in 1996, in an area designated as a priority district for smart growth under the state office. King Farm has a mix of housing types: 3200 dwelling units on 204 ha. The development also includes over 275,000 sq m of offices, and a village centre with 11,600 sq m of retail. The developers, Penrose Group, will build multifamily housing near the transit lines into DC. The project is taking shape quickly, in large part because of its transit access to Washington. An internal shuttle bus service provided by the residents' association facilitates access to the transit line and town centre (see Figure 4.4).

I visited King Farm on 27 April 2004 with a mobile workshop from the American Planning Association conference. The local planner told us that the zoning allows densities of 24 to 28 per ha. Single lots in the project are smaller than in some conventional developments, but still fairly large, at 465 to 740 sq m (5000 to 8000 sq ft). Detached homes in the development range from 325 to 465 sq m: large and

Figure 4.4 King Farm shuttle bus
Encouraging residents to leave their cars at home takes proactive planning. The residents' association at King Farm makes transit an easier choice by providing free shuttle services to residents.

expensive. Townhouses in the project have increased in value from $380,000 to $600,000 in three years, indicating the popularity of the development.

Despite rapid growth in the community, the commercial centre has proven a challenge for the developer (see Figure 4.5). It struggles to attract custom from the main artery that bypasses it. Several store fronts have windows with 'for lease' signs or advertisements for the project. The commercial area in the nearby Fallsgrove project resolved such problems by making a mall with interior parking lot its 'village center' (see Figure 4.6).

Figure 4.5 King Farm shops
Good quality materials and contrasting colours make the shopping district an attractive addition to the community. With its location near the transit service, the centre will likely improve its leasing ratio over the next few years.

Figure 4.6 Fallsgrove 'village center' mall
Shops at the Fallsgrove village mall line the streets but encircle the parking area. The 'village center' sign welcomes visitors to the parking lot.

These suburbs of the Washington DC area reflect regional patterns of segregation. While I encountered many women of colour pushing baby carriages on this warm spring day, most of the babies looked white.[4] The on-going construction of apartments near the rail line may change that picture and bring in more diverse and less affluent households, but it seems unlikely that the prevailing trends will change.

The 1990s also saw transit-oriented developments appear. These projects focused less on traditional design and more on compact mixed use near public transportation nodes. One of the better known developments is Orenco Station, about a 40-minute light rail ride from Portland, Oregon. This transit village, designed by Fletcher Farr Ayotte (2001), includes 800 living units, along with offices and retail on 77 ha (Bohl 2002). The developers, Pacific Realty Trust, placed a mixed use town centre near a main arterial road, some 400 m from the light rail station. Bohl (2002:15) reports that live-work townhouses in Orenco sold for $500,000, and that lofts 'have garnered prices that are unprecedented in Portland's outlying suburbs'.

When the fields between the station and the town centre fill, Orenco may live up to its promise as a medium density transit node. As of the fall of 2004, though, most of the homes and shops of Orenco Station lay a good 15-minute hike from the station; the Orenco Gardens project on the other side of the tracks is closer, but primarily residential. The town centre buildings feature two storeys over retail, with parking tucked behind. Only a few blocks from the arterial road, single detached houses on relatively small lots make up most of the site (see Figure 4.7). Despite the Portland region's commitment to new urbanist principles, making a project like Orenco Station work will take several years, and deep pockets.

Figure 4.7 Orenco Station house
Good transportation access to the city helps to sell homes in the Oregon countryside. Turning a transit node into an urban setting will take a few years.

Another TOD project, The Round in Beaverton (Oregon), took off slowly at first, but the *New Urban News* reported in March 2004 that by then the project was selling well. Many urban sites with good access to rail or subway connections enjoyed new prosperity and gentrification by the late 1990s, as the costs of traffic gridlock encouraged people to take a new look at city living. Also, federal investments in public transportation promised enhanced opportunities to consider transit-oriented design.

Interest in urban living as a model of smart growth gave impetus to many projects to find new uses for redundant industrial lands and malls. Many of these projects have employed new urbanist principles such as mixing commercial, office, and residential uses and looking to vernacular architecture for inspiration. Federal Realty developed several projects: Santana Row in San Jose (California), Bethesda Row in Bethesda (Maryland), and Pentagon Row in Arlington (Virginia). These projects renovated or rebuilt failing malls or commercial areas. Some developers experienced considerable success in the model (Springer 2000), but others have found financing a challenge. Santana Row struggled in its first year after a fire damaged construction, but is now doing well (see Box 4.2). Modelled after Las Ramblas in Barcelona, the project features high end stores, hotel, and housing in a Mediterranean theme (Perkins 2001).

In many growing cities, urban revitalization is transforming older urban areas. In the Pearl District in Portland, Oregon (Berton 2004), for instance, loft conversions happen quickly, changing the mix of uses in former industrial areas to include residential and commercial activities. With good access to the street car line, the Pearl District is experiencing rapid increases in property costs. Areas colonized by artists have now become popular for urban professionals. Some visitors to the area see it as a 'theme park': a comfortable lifestyle for the upwardly mobile (Berton 2004). Upscale furnishing retailers and chi-chi restaurants vie for the consumer's attention. Similar processes of gentrification are happening near transit lines in the inner cities of many high growth and high cost cities, like Washington, New York, and Chicago.

In the wake of the growing popularity of new urbanism and smart growth, some communities have been thinking about whether they can use new urbanism principles to retrofit the suburbs (e.g., MSM 1994). Creating town centres and urban villages to promote higher densities and animate suburban spaces has received new attention (Bohl 2002). To date, though, many areas have resisted intensification and diversification of the suburbs. Despite a fair level of consensus in the professions, and increasing interest in the political arena, consumers generally seem fairly happy with their suburbs as they are (Baxandall and Ewen 2000; Talen 2001).

NEW URBANISM IS BORN 93

Box 4.2 Santana Row: the mall becomes a village

Santana Row was conceived of as an urban mixed-use village that would serve as a regional retail and entertainment destination in the San Francisco Bay Area (see Figure 4.8). Opened in November 2002 in suburban San Jose, Santana Row is the largest mixed-use development in the Bay Area. Located on the 17 ha site of a former mall, when completed Santana Row will offer 6300 sq m of retail space, 1200 residential rental units ranging from lofts to townhouses, 5200 parking spaces, several green spaces and plazas, a central park, a 214-room four-star boutique hotel, and a multiplex cinema. The centrepiece of the development is the pedestrian-oriented, 457 m street flanked on either side by three storeys of residential space over ground-level retail.

Figure 4.8 Santana Row: the mall goes retro
In San Jose CA a failing mall site has been redeveloped as an upscale 'village centre' with a mix of commercial, residential. and entertainment uses.

Praised for its mixed use and new urban elements, as a grayfield development, Santana Row succeeds at intensifying and providing a greater variety of uses than the single-storey mall that preceded it. The grid street layout, building densities, and pedestrian pathways make Santana Row's amenities easily walkable within a few minutes, and the 40-year oak and palm trees saved from the previous mall site help make the walk pleasant. The streets are scaled for the pedestrian, with contiguous and narrow building frontages and wide, intermittently-shaded sidewalks. The streetscape is visually interesting with varied building heights and materials, a blend of architectural styles and landscape design elements.

On-street parking is permitted along some of the streets, but most parking is relegated to garages on the development's periphery. This parking is essential to retail success given Santana Row's suburban location at the intersection of two major arterials. An aerial view would reveal the lack of connectivity to the surrounding area and the vast area dedicated to parking. While bus lines run along the arterials, the site is not near a major public transportation centre. Although the development attempts to connect with adjacent low-density neighbourhoods by providing several entrances, the wide arterials and location amidst auto-oriented developments preclude much foot traffic to or from Santana Row.

As an upper income development, Santana Row is not a perfect example of new urbanism, despite its admirable urban design elements. Rents are above average for the area and there is no affordable housing component. The retail tenants are mostly upscale and echo those of Beverly Hills' Rodeo Drive. The development's amenities and residential marketing generally target young, upper-income professionals. Furthermore, Santana Row falls short in that it lacks civic uses, is not linked seamlessly to its surroundings, and is not truly a public space.

Like other large-scale projects by a single developer, Santana Row was built all at once (albeit in phases) and thus attempts to mimic incremental growth with its varied architectural styles. The developers, Federal Realty Investment Trust, hired several architectural firms to design the buildings and different artists to develop finer-scaled elements, such as fountains. The result is eclectic, though generally more reflective of European architecture than Bay Area vernacular. Ironically, a church façade was imported from Spain to be used for character in Santana Row's central park.

Overall, Santana Row offers architectural interest and an aesthetically pleasing environment. The streets are clean and bustling on a Saturday afternoon. The shop windows are inviting and the central park and streets offer opportunities for retailing, eating, relaxing, or making chance encounters. But regardless of how pleasant a walk through it can be, Santana Row has a fabricated, Las Vegas quality which results in a place that does not feel authentic. In many ways, a visit to Santana Row feels like a trip to an outdoor shopping centre, not the neighbourhood main street it attempts to emulate. Whether that will matter to Santana Row's visitors and residents, or if it will affect its longevity, remains to be seen.

Source: Sue Beazley

NEW URBANISM FOR POOR NEIGHBOURHOODS

One of the key arguments of the new urbanism in the USA involved the need to promote affordable housing. New urbanists have clearly stated, however, that addressing poverty must avoid concentrating low income housing. They saw the modernist public housing projects of the post-war period as clear failures. Concentrating poverty creates conditions for crime and decay. As Duany said in an interview, it is necessary 'to decant the monocultures of poverty' (Marshall 1995b:online). Since poverty becomes concentrated by modernist growth and the mechanisms to control it, such as exclusionary zoning (Downs 1999), the best way to overcome the problems are through planning in a new way. While a wide range of factors were responsible for the deplorable condition of American public housing by the 1990s, planning and design certainly came under criticism for contributing to poor living environments. New urbanism offered new approaches to dealing with the problems.

Duany et al. (2000) argue that a key problem of poor neighbourhoods involves the insufficient number of middle class people in these areas. In 1992, the Housing and Urban Development department announced a new programme, HOPE VI (Housing Opportunities for People Everywhere), with a mandate to demolish, refurbish, or renew public housing (Goetz 2003). Originally the programme required one-for-one replacement of lost units, but that requirement disappeared in 1995 (Swope 2001). The Clinton administration (1993–2000) proved anxious to improve the dire situation of public housing and increasingly sympathetic to new urbanism.[5] The government accepted the logic that concentrated poverty disadvantages people, and that mixing could improve the lot of the poor (Goetz 2003).

HOPE VI involved plans to demolish the worst 'barracks style' units, fix others, and improve management. Townhouses and garden style apartments would replace hated high-rises (Bohl 2000). A mix of market-rate and subsidized units would create more balanced communities. In many projects, only one-third of new units would be subsidized. Poor households displaced by demolition would receive 'Section 8' rent vouchers that they could use in arranging alternative housing.

Proponents of HOPE VI feel the programme has worked well. Steuteville (2003b, 2004b) reports that crime is down, while investment opportunities and incomes are up. Housing conditions have improved in many neighbourhoods, to the delight of local authorities and development interests (Popkin et al. 2004). Steuteville (2004b:2) says 'its success is a victory of common sense and of principles of urban design (as well as better community policing, tenant management, and the virtues of mixed-income neighborhoods)'. How do we separate out the factors to understand the role design plays?

Brophy and Smith (1997) argue that mixed income housing can work with good management and good maintenance, but good location also helps. Swope (2001) acknowledges that in projects like East Lake in Atlanta, crime is down and prices are up. He notes, though, that these new residents are not the same people who lived in 'the projects': only 79 of 428 families returned. Screening processes eliminated those with criminal records, drug use, or poor payment histories (Goetz 2003). Popkin et al. (2004) note that some poor families ended up worse off after HOPE VI than they were before. Many simply moved to new ghettos in the inner suburbs.

In early 2004 *New Urban News* reported that by January 2004, 193 HOPE VI grants had been issued, but only 26 projects completed. The Bush administration cut back on the funding, and was not ready to renew the programme.

Critics of HOPE VI wonder who benefited from the programme. Public housing sat on large pockets of consolidated land, often in good locations near core areas: selling off units or land in these areas proved good business for government. The gentrification that results from improvements allows new classes to move into well-located districts, and enhances tax revenues. If all the projects are completed, however, the USA could experience a net loss of 60,000 affordable units (Goetz 2003). Urban professionals interested in moving into the city benefited, but thousands of poor families lost access to affordable housing.

What happens to those displaced? Wexler (2001) says that most residents dislodged by HOPE VI opted to take the rent vouchers: they did not want to move back to the renovated neighbourhoods. However, some could not use the vouchers because affordable units were not available, or landlords proved reluctant to rent to them (Goetz 2003). Turnover in some of the projects has been very high (Brophy and Smith 1997).

Does social mixing work? Brophy and Smith (1997) find that residents in subsidized units mix more with each other than they do with market rate tenants: mixing can generate tensions between market and subsidized households, since they have different composition and behaviours. Although E. Wright (1997) sees less class consciousness in America than in Sweden, he notes different consumption behaviour and values, and few cross-class friendships. The polarization in society is very real and difficult to bridge: mixing housing units are not the same thing as social integration.

While HOPE VI has transformed some of the worst public housing into attractive neighbourhoods, it has also demonstrated that design cannot solve all the problems of the poor. If designers are not cautious, they may make things worse for some households. Design alone cannot change society.

NEW URBANISM GOES MAINSTREAM

Robert Davis founded the Seaside Institute in 1982 to study traditional neighbourhood development and educate people about it. The Congress for the New Urbanism (CNU) held its first annual congress in 1993, and ratified its Charter in 1996.

As early as 1989, Andres Duany was invited to give talks at conferences of the American Planning Association (APA), and seminars to the American Institute of Certified Planners. Professional journals like *Planning* profiled many new urbanism projects through the 1990s. By 2002, APA established a new urbanism division.

Developers and builders were also taking note. The Urban Land Institute (ULI) held seminars and began featuring many new urbanist projects in *Urban Land*. In

1998, the National Town Builders Association formed to help developers interested in new urbanism. Pattern books of traditional house designs appeared, and in 2001 the Architects Guild formed to provide better plans. During the 1990s, the CNU, APA, and ULI all began producing reports, books, and kits to support new urbanism approaches.

While many planners report that in their experience rank and file engineers generally oppose new urbanism, the Institute of Transportation Engineers has begun working with CNU and the US Environmental Protection Agency to develop alternative standards for streets. Langdon (2004a) says that although old road standards allowed engineers to reduce street requirements, road designers need more specific guidance to move forward with implementation.

By 1999, states like Maryland, Washington, and Oregon were leading the charge against sprawl and talking about 'smart growth'. In 2001, 27 state governors made smart growth proposals. While over 2000 bills were brought forward between 1999 and 2001, only 20 per cent passed (*New Urban News* 2002). Despite the growing professional consensus about the wisdom of smart growth, not everyone was ready for a change.

Many of the organizations supporting new urbanism put their weight behind smart growth. ULI held seminars, launched books, and included success stories in its magazine. APA developed a Growing Smart initiative, and by 2004, promoted form-based codes with the CNU (APA 2004). Form-based codes have the potential to institutionalize new urbanism within planning. Where developers imposed the design codes of earlier projects, form-based codes would replace or augment local government's permitted use system of zoning.

New urbanists have argued that codes are needed to ensure the quality of buildings because zoning and design guidelines are insufficient (Langdon 2004b). In an essay, 'Why write codes?' circulated by Emily Talen to the PLANET list on 15 May 2003, Andres Duany wrote: 'We must code because, if we do not, buildings are shaped by fire marshals, civil engineers, poverty advocates, market experts, the accessibility police, materials suppliers and liability attorneys.' Bureaucrats and professionals understand and will enforce codes. Codes can protect the designer's vision.

Sarasota, Florida, adopted the SmartCode from DPZ in May 2004, thus implementing the transect concept (*New Urban News* 2004a). Form-based codes adopted in Petaluma (California) and Arlington County (Virginia) have enabled planners to streamline responses to developers down to six months (*New Urban News* 2004a). Langdon (2003a) notes that places like Belmont (North Carolina), Chattanooga (Tennessee), Fort Collins (Colorado), Hillsboro (Oregon), Pasadena (California), and Seattle (Washington) have implemented zoning codes promoting TND or design guidelines.

Despite the obvious successes of new urbanism in influencing the way that Americans think about the city and suburbs, and the extensive lobbying of organizations like the CNU, ULI, and APA, the long-term viability of new urbanism as a planning approach in the USA is not fully secured. The model faces significant challenges in practice.

THE CHALLENGES TO NEW URBANISM

One of the key challenges to new urbanism is the security issue. In the wake of the terror attacks of 9/11/2001, protecting public buildings from potential terrorist attacks has focused attention on design issues. New security standards make it hard to design public buildings with mixed use, active street frontage, and underground parking. Langdon (2004c) worries that this may force designers to create hard and blank edges with a potential loss of urban vitality.

Fire departments have not been convinced of the appropriateness of narrow streets. Emergency equipment keeps getting bigger, making access an issue. Some people have raised concerns about whether new urbanist projects limit access for the mobility impaired. The front steps to porches of traditional-style buildings are inaccessible to wheelchairs. Advocates for the disabled want to require standard at-grade or street-level entrances for homes and businesses, and an accessible bathroom at ground-floor level, to enable the disabled to visit anywhere. Designers have suggested constructing rear or side entrances at grade, while elevating front entrances for privacy, but that may not always prove feasible. 'Visitability worries some designers because if it becomes a rigid dictate, it could make certain kinds of housing almost impossible to build, with damaging consequences for New Urbanism' (Langdon 2002a:4). Groups for the disabled have been protesting at CNU congresses each year since at least 2001 (Langdon 2002a).

The challenge of producing viable retail districts represents one of the weaker links in American new urbanism. Marshall (2000:11) says that: 'Retail is an area where fictions are exposed.' Most new urban projects have not established vibrant commercial centres. New urbanists often suggest that 1500 new housing units can support one block of stores, but with 10 per cent vacancy in commercial properties in the USA, Bohl (2002) confirms that retail supply clearly exceeds demand. Many new urbanist projects have empty storefronts, sometimes waiting years for retail to appear, or reducing rents to attract tenants. In cases where stores fill, as in Celebration, the mix may not appeal to local residents. Celebration residents want a hardware store, not gift shops. While the doctor's offices and real estate businesses in Kentlands may generate some custom, they do not animate the street. The village centre mall of a place like Fallsgrove hardly substitutes for an old-fashioned main street. In some of the town centre projects, like Santana Row, entertainment and high end retail uses dominate: the retail is viable but may displace sales elsewhere in the city. American retail patterns that favour big-box shops, chain retailers, and mega-malls threaten the commercial ideals of new urbanism (Langdon 2003b).

Managing commercial traffic remains an issue in new urbanism. Early projects located the retail centre away from arterial roads to create civic hubs. Out of sight of non-residents, the businesses could not attract sufficient volume to thrive. In response to the financial difficulties, many new urbanists now advocate placing commercial uses on arterial roads at the edge of the neighbourhood: that entails a compromise in walkability, since the retail will no longer be central to all homes. In effect, the adjustment replicates some features of Perry's (1929) concept of the neighbourhood unit which placed commercial uses and apartments on the bordering main roads.

Some might ask whether new urbanism represents a new mass form or an élitist option. The principles of new urbanism favour supplying affordable housing, but new urbanists often advise investors that new urban projects appreciate more quickly than other projects and enjoy a market premium (*New Urban News* 2003b; Song and Knaap 2003). Tu and Eppli (1999) found that Kentlands homes were $25,000 more expensive than homes in comparable conventional developments. Eppli and Tu (1999) suggest that Harbor Town in Memphis has a $30,000 premium over the competition. Katz (2002) says that Rosemary Beach (Walton County, Florida) saw an annual increase of 32.8 per cent in values (versus 8.3 per cent for a comparable nearby conventional development); I'On in Mount Pleasant, South Carolina, appreciated 37.5 per cent in one year, versus 5.9 per cent in a conventional equivalent. In sum, good new urbanism increases property values.

Prices in new urbanist projects are usually high, affecting the social composition of the resulting communities. For instance, in Aqua, Florida, townhouses and medium-rise condominiums in a low density development sell from $500,000 to $7 million (Contreras and Villano 2002). Nothing is available at Seaside for under $600,000. Even if projects begin as affordable, they seldom stay that way.

How strong is the commitment to affordability? Alex Marshall (1995c:online) quotes Duany as saying that 'affordable housing is not what cities need. Because they don't pay taxes. They bankrupt cities'. If cities need the middle class to come back, then new urbanism can contribute to improvement. But if most new urbanist projects remain in the suburbs, then they only exacerbate urban problems.

Despite the billions put into HOPE VI, we find few cosy American communities of poor households in traditional homes around public squares or over community shops. Most new urbanist projects are high-end, with little affordable housing. Projects built in formerly poor districts have often moved poor households out. New urbanists see this as urban 'evolution', but others call it 'gentrification'. Marshall

(1995a, 1995b) goes further, describing the agenda as a new kind of urban renewal: a readiness to destroy neighbourhoods to impose physical solutions to social problems. In the East Ocean View project in Norfolk, Virginia, Marshall says that Duany would reduce 1800 affordable homes to less than half that many. Half-million dollar homes would replace smaller cottages on the beach. 'The New Urban idea of diversity is to sprinkle a handful of middle-class residents within a solidly upper-income subdivision' (Marshall 2000:28). Does new urbanism help to drive the poor and working classes out of neighbourhoods? Does the good community have no poor people because their status has improved, or because they are forced elsewhere?

The new urbanists hope to change the tone of behaviour in the American city. For example, in a report on Providence, Rhode Island, the *New Urban News* (2003a:12) reported that 'downtown Providence has a conspicuous number of coarse people'. Mixing the poor with the middle class, and ensuring community policing and surveillance, are presumed to make areas attractive for investment and redevelopment. Kunstler and Duany see the suburbs as contributing to bad behaviour among youth; they hope that better community design will improve the situation. But even as we sat in the main town green in Kentlands on a lovely Saturday evening in early October 2004, we were joined by four teenagers dressed in black, swearing up a storm, and drinking from wine bottles hidden in paper bags. Design alone cannot civilize behaviour.

We find a similar problem with travel patterns. 'The basis for using land use and urban design to selectively change travel behavior ... appears limited in the near term' (Crane 2000:3). In most new urbanist projects, residents remain car-dependent for work and shopping. Some projects have good access to mass transit, but most do not. As Talen (2001) notes, suburban residents are perfectly happy using their cars. They may agree with some new urbanist concepts, but that is not enough to change their behaviour. As Audirac *et al.* (1990) and Southworth (1997) suggest, policies that go against cultural values prove extremely difficult to implement.

Some of the most vocal critics of new urbanism and smart growth, such as Wendell Cox (2004) and Randal O'Toole (2003, 2004), go further. They see the movements as reducing choice and ignoring the preferences of American consumers. Efforts to promote new urbanism are likely to fail, they believe, because the movements contravene market logic and ignore individual liberties.

New urbanism also faces challenges to its environmental agenda and approach. New urban projects often set aside large green spaces and parks, but designers readily manipulate the environment. Many projects have created lakes and canals to replace wetlands, and covered farmlands and forests with houses. Frantz and Collins (1999) criticize the Disney staff for paying little attention to the natural environment in Celebration: developers removed every possible tree to squeeze in more houses. With the popularity of the transect increasing, new urbanism encourages designers to avoid overly naturalistic green areas in urban environments. 'We must code in order to assure that urban places can be truly urban and that rural places remain truly rural, and that there be a specific transect in between. Otherwise, misconceived environmentalism tends to the partial greening of all places', says Duany (Talen 2003:online). New urbanists see no virtue in greening for the sake of greening.

Projects designed for water edges reflect the potential conflict between protecting fragile ecosystems and achieving vibrant urban settings. Steuteville (2004c:9) reports on a proposed development in Florida that promotes 'forthright urbanism, which in the center of town marches right up to the water'. He criticizes a North Carolina state policy that calls for a 9 m buffer for waterfront areas that he believes will 'thwart the creation of compelling urban environments at the water's edge' (2004c:9). Images of the project in his article show boathouses of private residences hugging a hardened shoreline. Ecologists would surely be distressed. New urbanists seldom discuss habitat protection or landscape function in their pursuit of good design. Audirac et al. (1990) worry that the compact growth policies adopted by states like Florida may worsen environmental problems, like threats to sensitive wetlands. If intensification occurs, central areas set aside for environmental protection may be converted to urban uses, thus depriving urban residents of ready access to natural sites. While conserving environmental lands outside the city is essential to long-term sustainability, critics find the new urbanist approach to the natural environment within urban and suburban environments short-sighted.

THE IMPACT OF NEW URBANISM IN AMERICA

We cannot deny the impact of new urbanism on urban and suburban form in the USA. While the number of comprehensive projects remains small, some traditional design elements are widely borrowed in the mass market. 'Retro' architectural features like porches and dormers have gained great popularity. Window inserts and façade detailing often emulate historic styles. In many cases, unfortunately, we have to conclude that developers are stealing the look without the content (Ehrenhalt 1997; Grant 2002a).

Both Banai (1998) and Fader (2000) report that lots are getting smaller in new urbanist projects, with 230 sq m (2500 sq ft) lots appearing. Given that lot sizes of around 930 sq m became common in the post-1945 period, this is a clear step forward (Banai 1998). Mixing lot sizes and housing types has also become more acceptable. Houses are moving forward on lots, and streets are narrower. These changes reveal the transformative capacity of new urbanism in defining good community form.

The question, then, is whether new urbanism can stem the tide of urban expansion. New urbanists may justly argue that until the approach becomes standard, it will be difficult to demonstrate the savings possible. Conventional development patterns prevail in the suburbs, consuming vast amounts of land each year. Toni Alexander (2000) says that the American LIVES survey shows that the complete new urbanist package appeals only to about 10 per cent of population. Most homebuyers want larger lots and low density neighbourhoods. While 75 per cent say they want to live in a town, only 49 per cent prefer small town community traits. The bottom line is that few households prove willing to give up their cars or big front yards. Two decades of demonstration projects and extensive media coverage have not convinced the general public to buy the new urban community model, although Ellis (2002) believes that change is coming. Moreover, even if lot sizes and household sizes may be decreasing, houses are getting larger (Hayden 2000): in 1950, the average size of a new house was 74 sq m; by 1970, 139 sq m; in 1998, 203 sq m (2190 sq ft). Meanwhile, household size declined from 3.38 in 1950 to 2.59 in 2000 (US Census 2002). The average household buying the average new house in 1950 had 21.89 sq m of house per person; the average household in the average new home in 2000 had 78.38 square metres per person. Fewer people are consuming more housing, with all the materials that involves. The pattern is not sustainable and not smart.

Marshall (2004) argues that new urbanist projects are just suburbs in disguise. They remain car-dependent. They consume land at the urban edge without becoming urban places. They divert attention from pressing problems with 'this pretence that American society can build its way out of the problems of suburban sprawl' (Marshall 2004:online).

In some places where new urbanist projects have led to rapid growth, a backlash against higher densities has resulted. The *New Urban News* (2004b) reported that Kyle, Texas, would increase its minimum lot size because of a local perception that the Plum Creek new urbanist project was too dense. Where the quality is high and demand for housing of any kind relatively strong, then new urbanist projects prove successful and are emulated. In areas where land costs remain relatively low, they may struggle to make headway in the market. In many districts, they remain strictly illegal under local land regulations.

Despite the good press for new urbanist projects, and bad press for gated communities (e.g., Atkins 2003; Barstow 2001; Chiotti 1992), secured enclaves are appearing with greater frequency than new urbanist communities. As Blakely and Snyder (1997) note, Americans are 'forting up'. Sanchez and Lang (2002) report that some four million households live in tens of thousands of access-controlled communities. By contrast, fewer than 400 new urbanist projects were built by 2004, containing perhaps 300,000 households. New urbanism has captured a small share of the market.

Despite the slow progress and a cadre of critics, the proponents believe that new urbanism can create vibrant new neighbourhoods in suburban areas that otherwise may give people little cause to interact. The CNU congresses attract more people each year, and professional bodies and lobby groups continue to press the message vigorously. Some university architecture and planning programmes now teach new urbanism not as one among many competing ideas, but as the vision for a better future. The movement is still growing, and has been consolidating its influence.

What are the next steps for the American movement? The CNU debated the idea of a certification programme for new urbanist practitioners. The APA is promoting form-based codes with vigour. ULI advocates mixed use, higher density projects to its members. Other organizations work assiduously to promote new urbanist methods, solutions, and techniques, from traditional architecture to design charrettes. The movement shows few signs of weakening just yet, even though it faces significant challenges to its hopes of changing suburban form and redefining the shape of the good community in America.

CHAPTER 5

REVAMPING URBANISM: THE EUROPEAN EXPERIENCE

A HISTORY OF URBANISM

Europe has a vibrant urban legacy, yet its cities also face the crises of the contemporary era. Issues of sprawl, crime, and traffic force residents to consider alternative strategies to accommodate the needs of new households. In this chapter, we consider how new urbanism may be influencing planning practice in Europe. Since the Renaissance, proponents of classical forms have always had a voice in Europe. Revivals of traditional architectural practices occurred several times in recent centuries. To what extent are new urban approaches indigenous to Europe, and to what degree do they respond to the new urbanism movement in America? Many of the elements of new urbanism draw on models of the European city, so we should not find it surprising to see the movement gain popularity abroad. At the same time, though, Europeans may resist adopting ideas from 'the colonies'. While a few architects are working diligently to establish a European (new) urbanism movement, by and large we have to conclude that European nations are making their own new urban approaches that are similar in many ways to trends in North America, but with some unique features. The European image of good community form does not prove entirely congruent with the American new urbanism prescription.

The traditional European city has become iconic in Western culture: Paris as the city of love; the beautiful canals and piazzas of Venice; the fine residential squares of London; the dynamic intensity of Barcelona. Architects-in-training make pilgrimages to many of the great cities for inspiration. For North Americans, raised in cities with short histories, or in suburbs with little character, the European city presents a vibrant example of what is possible and desirable.

Of course, European cities have changed considerably through time. An urban culture first came to southern Europe with the Myceneans and Minoans in the sixteenth to thirteenth century BC (Chadwick 1976; A. Morris 1994). Following a period of decline and de-urbanization from the twelfth to ninth centuries BC, urban development resumed in Greece with the development of the *polis*, or city state. The Greeks invested phenomenal resources in their infrastructure and public spaces, building fine stone temples, fountains, offices, and meeting places for state activity (Owens 1991). By contrast, the private living quarters of the city featured poor mud-walled dwellings (A. Morris 1994). After the fifth century BC, the Greeks increasingly used grid layouts for their planned cities, a practice perfected by their successors in the power struggle for the Mediterranean: the Romans (Owens 1991). Thus throughout the region, the city signified the coming of culture as the great civilizations colonized the land by planting cities to control or take territory.

During the fourth and third centuries BC, the Romans established dominance over Greek and Etruscan city states and began to develop a great empire, knitted together by a chain of colonial towns and cities. Adopting and refining the urban traditions of the Greeks and Etruscans, the Romans established a fine capital at Rome, investing the wealth of the empire in public buildings, streets, sanitary networks, and statuary. The Romans spared no expense in beautifying their cities, including their private homes and villas, perfecting classical principles of design in the process.

To hold the vast territories that fed the empire, the Romans built military and trade towns, typically according to a set plan that varied little from region to region (D. Hughes 1994; Owens 1991). Town governors vied with each other to create the most beautiful public buildings and spaces. The Romans thus brought the first flush of large-scale urbanism to northern and western Europe. Some sites of Roman occupation remained urban centres through the centuries. For instance, York in northern England built on its Roman roots through subsequent occupations by Angles, Saxons, Norse, and Normans. Some cities, like Paris and London, were abandoned as the Romans withdrew with the decline of the empire in the fourth and fifth centuries, but regained their significance as trade grew again during the middle ages. Some Roman settlements, like Calleva Atrebatum (Silchester, UK) returned to fields and forest, their purpose lost with the empire that created them (Roman Britain Organization 2003).

In southern Europe, the Byzantine Empire and subsequent Arab empires continued vibrant urban traditions with new design principles inspired by different cultural motivations than shown in the classical Greek and Roman cities. Cities like Marseilles, Constantinople (Istanbul), and Venice thrived on Mediterranean trade. Cut off by hostile tribes in central Europe, and Islamic empires in the south, northern Europe reverted to a subsistence economy. An urban network gradually began reviving as Europe was pacified following widespread conversion to Christianity around AD 1000, and reconnected to the Mediterranean as Arab powers were pushed out of southern Europe.

Growth in trade in wool, textiles, and slaves fuelled European urban development in the Middle Ages (Pirenne 1956 [1925]). Centres of textile production and trade, such as Bruges and Ghent, grew rapidly (Figure 5.1). Enclosed within walls and moats for security, the medieval city was typically compact, with little open space except for markets and church squares (A. Morris 1994). Its unplanned street network and limited size inspired garden city advocates like Lewis Mumford (1961) to identify the medieval town as a model of good urban form.

Figure 5.1 The medieval city as the good community
Bruges (Belgium) reached the peak of its influence and splendour in medieval times. Garden city advocates saw such compact places as excellent examples of good urban form.

In the wake of depopulation from the plagues, and with new ideas deriving from the Enlightenment, European cities began to open up, with parks and squares and increasingly grandiose architecture. Classical principles made a revival as economic development and the consolidation of political power led to urban growth. Grand boulevards, monuments, and statuary allowed rulers to demonstrate their power. Architectural elaboration and decoration created beautiful capitals: the classic European cities such as Rome, Florence, Paris, and Lisbon (Girouard 1985).

With the rise of an industrial economy in the eighteenth and nineteenth centuries, new cities began to develop, usually on navigable waterways near the resources needed for production. Railroads linked them to resources and to markets. With rapid growth in cities like Manchester and Dresden, the focus turned to the efficiency and utility of buildings, rather than ornamentation. In periods of peace older cities tore down their defensive walls and filled their moats to create new boulevards or parks. The cities that became industrial powerhouses replaced old structures with new ones, transforming and modernizing themselves in the process. Cities by-passed by industry fell into decline.

In the twentieth century, much of the new building in European cities proceeded by the same principles as operated in North America: efficiency, simplicity, and economy. Modernism proved a powerful and persuasive paradigm for post-Second World War city centre and suburban development. The stronger the local economy, the greater the loss of pre-twentieth-century heritage.

In the post-industrial or new economy, cities function as centres of culture, knowledge, and power. This creates a second chance for older cities overlooked in the last half century's growth, such as Glasgow and Dublin: their historic buildings attract high-tech industries looking for heritage and character. Some former industrial giants, like Liverpool and Birmingham, declined as people moved elsewhere. Eastern Block cities integrated into new democratic nations have sometimes lost residents who emigrated to search for work. Thus the last two millennia reveal diverging fortunes for European cities in each period, with growth for some, and decline for others, depending on economic and political circumstances. A varied history has created a wealth of urban resources: great buildings, fascinating neighbourhoods, and beautiful parks.

Important personages in history treated the city as a canvas on which to inscribe their legacy: Popes built fountains and fine statues in Rome; Louis Napoleon (Napoleon III) fashioned the grand boulevards of Paris; the Prince Regent left parks and parades in London. With the rise of the nation state in the nineteenth century, however, urban architecture lost its greatest patrons. The wealthy barons of industry and commerce sponsored some projects, but not to the extent of their aristocratic predecessors. Some buildings, such as the Casa Mila in Barcelona (by Antoni Gaudi), and the Guggenheim Museum in Bilbao (by Frank Gehry), warrant mention, but in the last century Europeans built mostly unimpressive buildings. A long legacy of development, followed by a recent spate of uninspiring construction, has created significant challenges for European cities.

Challenges of the contemporary European city

Some of the problems facing European cities derive from the rich tradition of urban development in the region. While many older buildings survive, they often need renovation and expensive maintenance to remain functional. Pressure for land near the city centre renders older urban areas vulnerable to change. Growing demand for intensification of use may make land so valuable as to threaten heritage resources. For much of the nineteenth and twentieth centuries, business and government thought it desirable to replace heritage buildings with modern structures. As a consequence, in some areas unique regional styles and ancient building skills began disappearing. In the late twentieth century, however, renewed interest in heritage conservation helped to generate a market for revitalizing older areas and finding new uses for historic buildings. Such revitalization contributes to tourism and economic expansion, but also can fuel inflation and drive poorer households and small businesses away from the city centre.

Central cities in Europe, with their narrow, winding streets and their four- to six-storey buildings, feel quite different from North American cities. They often feature peripheral 'slab-urbs': high density, pre-fabricated apartment blocks (Eurocouncil 2003). Some critics suggest, though, that suburban areas are becoming more like American suburbs all the time, with separated land uses and urban functions (Angotti 2002; CEU 2003; Eurocouncil 2003).

As populations become more affluent, car ownership has grown steadily, increasing traffic congestion (European Environment Agency 2001; UK National Statistics 2002). Traffic has had a significant impact on the road system, especially in cities with medieval streets. Some cities have turned central streets into pedestrian precincts or one-way zones (Gehl and Gemzoe 2003). Inevitably, though, air quality has suffered due to exhaust emissions, and traffic safety has become a problem. The average commuting distance and time is increasing, with UK residents travelling the longest in Europe: 45 minutes per day (BBC News 2003). Across Europe, the rise in consumerism and individualism is transforming the landscape: people want more housing, more land, and more cars.

With cities having installed key central city infrastructure of pipes, drains, roads, and bridges in previous centuries, the cost of maintenance and renewal to bring systems up to contemporary standards can prove staggering. In some areas, demographic and economic change have rendered the urban infrastructure of villages and towns redundant. In the areas within commuting distance of large cities like London and Paris, the countryside faces suburbanization. British villages are integrated into the urban system as they are increasingly colonized by urbanites looking for the rural experience. Such villages may house primary residences for commuters or weekend retreats for the affluent. Local people can barely afford the costs of remaining as their taxes rise and house prices sky-rocket. For example, in July 2004 a thatched cottage in Pitsford (north of Northampton) was selling for over £300,000: residents expected someone in London might buy it as a country home. The mix of uses that once characterized villages like Pitsford has disappeared during the last few decades. Village retail is endangered. Thirty years ago Pitsford had two chapels, a hospital, a hotel, an almshouse, several shops, and three pubs. By July 2004, it had a single pub left; all the other buildings had been renovated for housing. To buy a loaf of bread, villagers must drive the seven minutes to town. The same loss of village stores is happening in other parts of Europe as shopping patterns change.

Thousands of years of intensive agricultural and industrial uses in Europe have had a devastating impact on the environment (Simmons 1993, 1996). Forests have disappeared. Waterways have been polluted, redirected, and in some cases filled. In the twentieth century, changing agricultural practices began to transform the traditional countryside, as hedgerows disappeared and factory farming displaced family farms.

At the same time as economic and environmental change threatens the countryside, though, old country names are reappearing. For instance, the village of

Hampstead Norris in Berkshire, where I lived as an infant while my father worked in the local bakery (since converted into a house), has changed its name to an older-looking spelling: Hampstead Norreys. New chains of pubs are built to resemble old-fashioned pubs. Packaged heritage replaces the authentic.

Many European cities have experienced a dramatic decline in the affordability of housing. In some countries, governments have reduced their involvement in social housing. For instance, the British government sold off much of its council housing stock in the 1980s and 1990s. The fall of communism in eastern Europe undermined the commitment to social housing there. More and more households search for private housing in an inflationary market. While private home-ownership rates are increasing, so is homelessness and overcrowding.

Cities that a century ago were relatively homogeneous in population, today face increasing social segregation by class, ethnicity, and age. Immigration is transforming communities; for some Europeans, this has increased fear of crime, and perceptions of difference and loss of tradition. Such conditions can make urban movements that look to the past for inspiration appealing. In light of the challenges threatening European cities, movements have developed calling for planning reform and new approaches to urbanism (Council of European Urbanism 2003).

European interest in new urbanism

While the American new urbanism movement unified under the banner of the Congress for the New Urbanism (CNU), European urbanists have developed several organizations promoting urbanism. With many urban traditions in different nations, Europe has a wealth of history and cultural difference to draw on in looking for appropriate urban approaches in the contemporary period. Since the late nineteenth century, some urbanists, like Camillo Sitte (1965) had argued for artistic approaches to urban design. Traditional architectural forms based on vernacular styles often enjoyed revivals. Gothic architecture experienced a florescence in the mid-nineteenth century before the Queen Anne revival and classical revival movements gained adherents (Rubinstein 1974). Europe also gave rise to the garden city movement and to modernism, a radically innovative urban approach. Following the Second World War, new development in Europe was typically modernist.

In recent years, several European countries and many urbanists have proven receptive to new urbanism, but we do not find a unified urbanist policy in Europe (Eurocouncil 2003). European governments have increasingly focused on historical preservation for tourism, economic development, and sustainable development since the 1960s (Council of Europe 2004). As the role of tourism in the economy grew in the late twentieth century, protecting and enhancing traditional urban districts became important. Heritage architecture appealed to professionals interested in urban living, stimulating new markets for residential development. The reaction against sprawl and ugliness in the urban environment encouraged many designers to look for alternative models of urban development.

Planning history reveals a long tradition of sharing of town planning ideas and expertise between North America and Europe. The garden city model reflected a transatlantic collaboration with Americans building on European ideas, and vice versa. The major players – like Raymond Unwin, Thomas Adams, Ebenezer Howard, and Lewis Mumford – travelled back and forth across the ocean, refining their ideas and sharing experiences. For the most part, though, experimentation with new town building happened primarily in Europe. Europeans might have concluded that garden city ideas were better implemented in Europe than in North America. Where the serious problems of American cities made Americans question the garden city modernist paradigm, Europeans instead explained the failure as lying with ineffective planning and weak political will on the west side of the ocean.

The sharing of planning ideas continued throughout the twentieth century, facilitated by new technologies of transportation and communication. As new approaches to urban development became popular in the USA in the 1980s and 1990s, the message about traditional urbanism made the eastward trip. While it certainly seemed logical that the Americans might look to Europeans for inspiration in reviving strong urban traditions, it was less clear that Europeans might see American advice as useful.

That said, of course, some of the key players contributing to new urbanist ideas come from Europe. One of the most important thinkers in the new urbanism movement is Leon Krier, the classicist architect from Luxembourg who inspired Duany and Plater-Zyberk and secured the patronage of the Prince of Wales for his projects. Robertson (1984) describes Krier as a non-building architect interested in planning cities. Krier's critiques of modernism struck a chord with many. In 1977, Krier wrote that the city needs urban quarters with multiple functions, and clear centres and edges. He slammed the suburbs as dead, cancerous, parasitic, and self-loathing places. As an alternative, Krier advocated following traditional principles for urban form, based on ideas Sitte (1965) had about the public realm.

As Thompson-Fawcett (1998) notes, Krier is committed to the rationalist search for common principles of design. He sees certain principles as timeless and universal. Form cannot just follow function or it creates isolated objects (Krier 1984a). The way to save the city is to abolish zoning, banish motorways, and clearly separate the city from the countryside. In sum, the good community requires that we abandon the principles and practices of modernism.

As Krier was writing and lecturing about the need for a return to classical principles, the Prince of Wales began criticizing the destruction of British heritage (Prince of Wales 1989, 2004). He called for architecture that respected the past, favouring 'some of the basic principles that have governed architecture since the Greeks' (1989:76). He decried the 'vast and neglected soulless deserts of post-war housing' (1989:87), challenging architects and developers to aspire to better cities.

The pairing of like-thinkers, Krier and the Prince of Wales, gave the architect a patron and the Prince a visionary designer. In 1989 they collaborated in creating the Urban Villages Group, and set about developing a concept for a town created by traditional urban principles. Poundbury village began construction in 1993.

The Prince also consulted American new urbanists like Duany, strengthening the links between American and European urbanism. He hosted architectural institutes and became a patron to organizations such as the Phoenix Trust, founded to assist with the restoration and reuse of heritage buildings. Eventually the trusts and institutes combined in the Prince's Foundation with a mission of promoting education and projects related to traditional urban development patterns.

Connections between North America and Europe have become extensive in recent years. Teaching in America in the 1970s, Krier inspired Duany and Plater-Zyberk to become traditional town planners. Duany and Plater-Zyberk's success at Seaside in turn influenced Krier and others in Europe to recognize the possibilities of the upstart movement. A small but influential contingent of designers travel regularly across the Atlantic, meeting at congresses and conferences, and enjoying invitations to teach or deliver charrettes (Ouellet 2004). Key players involved in the European new urbanism movement include Lucien Steil, Gabriele Tagliaventi, Joanna Alimanestianu, and Maurice Culot. American new urbanists like Andres Duany, Victor Dover, and Stefan Polyzoides regularly attend Eurocouncil meetings, and may collaborate with European partners on design projects. A strong network has developed.

In Europe, new urban approaches have built on other critiques of modernism. Through the 1980s and 1990s, the idea of the compact city proved popular in European thinking about planning. The compact city theory suggested that containing growth in a dense form could offer a more sustainable option than continued sprawl (Jenks *et al.* 1996). It reinforced traditional understandings of the European urban form. The new urbanist movement accepted the logic of the compact city and offered useful design options to help achieve it.

In recent years, governments in the UK have made a significant commitment to some of the principles of new urbanism. In 1999, the Urban Task Force, led by the architect, Lord Rogers, called for an 'urban renaissance'. The Urban Task Force report linked the language of sustainability and the compact city explicitly with ideas of traditional urbanism. The report called for compact form, mixed use, and integrated public transport.

> Excellence in the design of buildings and spaces cannot exist in isolation from a clear understanding of what makes for the most sustainable urban form. In this report we argue that the compact, many-centred city of mixed uses which favours walking, cycling and public transport, is the most sustainable form.
>
> (Urban Task Force 1999:40)

The urban renaissance quickly became a significant ideal in English planning. The Office of the Deputy Prime Minister (ODPM) certainly climbed on board. English planning guidance notes have been revised to reflect the principles (ODPM 2004a).[1] Projects to promote sustainable development and to construct urban villages are underway in many parts of England, and Scotland has also experimented with urban villages (Hague 1997, 2001). The Deputy Prime Minister, John Prescott, toured new urbanist projects and met with Andres Duany and John Nordquist of the CNU in the USA in 2003 (Steuteville 2003a). The renaissance seems explicitly to involve many new urbanist ideas.

Several other European countries are also committing to compact development options and urban design ideas that new urbanism promotes. The Netherlands has a strong movement for compact form to protect the 'green heart' of agricultural land from urban development (De Roo 2004). Redevelopment of brownfield lands has high priority, for instance, in Germany; Wiegandt (2004) suggests that new urbanist and smart growth ideas are affecting such development options. In many cases, though, it may prove difficult to differentiate business as usual in European development from 'new urban approaches'. After all, conventional development in many countries has tended to include fairly high densities, transit-oriented approaches, and a reasonable degree of mixed use.

Although we can see considerable evidence of the influence of new urban ideas in Europe, the movement is far from united. Several organizations, sometimes with overlapping membership, are pushing for new urban approaches in Europe (see Table 5.1). These groups operate in different cities, often with similar ends and limited means. Efforts to unite them under a single Council of European Urbanism (CEU) are proceeding, but have not fully succeeded. At recent meetings, the CEU (2003, 2004) has been working on a European charter modelled after the CNU charter.

Table 5.1 European organizations promoting traditional urbanism/new urban approaches

Council for European Urbanism	The Prince of Wales' Urban Design Task Force
Eurocouncil	
Foundation for Urban Renewal	The Prince's Foundation
Byens Fornyelse	The Prince of Wales' Phoenix Trust
A Vision of Europe	International Network for Traditional Building,
Urban Villages Group/Network/Forum	Architecture, and Urbanism (INTBAU)

While the 'European Traditional Urbanists' (Eurocouncil 2004) continue looking for ways to promote new urbanism in Europe, Ouellet (2004) suggests that national differences may limit efforts to unify the movement: the proliferation of groups diverts energy from disseminating a clear message.

Several university programmes are now promoting new urbanism: for example, the University of Ferrara (Gabriele Tagliaventi) and the New School of Viseu at the Universidade Catolica Portuguesa (Jose Cornelio da Silva, Lucien Steil) are collaborating with American schools teaching new urbanism, like the University of Notre Dame.

Although new urbanism is growing in influence in Europe, it is not universally admired. Bodenschatz (2002) suggests that Europeans see it as a 'provocation'. Europeans criticize the 'Americanization' of the city, with motorways everywhere, architectural chaos, and diminishing law and order. At the same time, though, they may see the traditionalism of new urbanism as 'backward-looking and a falsification of history' (Bodenschatz 2002:100). Modernist architecture does not have the negative connotations for Europeans that it has for many American designers. Certainly we see less focus on traditional architectural models in urban village projects in Europe than is true in the USA. New urban approaches in Europe often involve modernist architecture.

Debates about the form of the good city are nothing new: they are as old as the profession of town planning. And although we find little evidence of consensus, we can see that some new projects employ principles consistent with new urbanism. While their developers don't often use the term 'new urbanism' to describe what they do, these projects reflect the widespread currency of key design and planning principles in the contemporary period.

EUROPEAN NEW URBAN PROJECTS

How do European theory and practice compare with that in America? What kinds of new urban projects do we see being built in Europe? What new urbanism principles are practised and which are not? We hope to answer these questions in a brief review of some European projects. The review is by no means comprehensive or detailed, but it gives an idea of the international applicability (and limitations) of new urban ideas.

With its commitment to an urban renaissance, the English government has issued planning guidance indicating its desire to contain 60 per cent of new growth within already developed areas. Can it achieve this target? If government is successful, it will certainly increase urban densities; it also aims to reduce car use and land consumption. Home builders like the giant Wimpey say the policy will slow growth and force home prices up (Aldrick 2004); this in a market that is already extremely expensive. Greenfield development remains a significant component of new development in the UK. Even some of the urban village projects, like Upton Village south west of Northampton, are planned for greenfield sites (see Figure 5.2). The aim of the target is laudatory, but meeting it is likely to prove a real challenge.

Infill development on brownfield and grayfield lands is making an important contribution to urban revitalization in many European cities. Some of those projects follow new urbanist principles. The Rue de Laeken project, called 'famous' by the Eurocouncil (2004), reconstructed a block in Brussels in the mid-1990s. Developer Christian Lasserre and architectural advisor Joanna Alimanestianu set a block of four- to six-storey buildings around a private central courtyard. The block faithfully recreates the architectural feeling of an area of the city that otherwise seems a bit down on its luck (see Figure 5.3). When I visited the site in July 2004, the project had more than half a dozen empty store fronts on the Rue de Laeken and the Rue de Cirque. By contrast, the upper apartments and offices seemed fully occupied. The design of the project certainly merits praise, but perhaps that was not enough to reanimate a depressed neighbourhood.

Figure 5.2 Upton Village site plan
This greenfield urban village south of Northampton UK promises a compact layout in a walkable community. Despite government's commitment to contain 60 per cent of new growth within urban areas, these homes are displacing farmers' fields.

Figure 5.3 Rue de Laeken: new traditions
The Rue de Laeken infill project in Brussels captures the design principles of a district that has seen better days. Using styles that reflect architectural heritage has become a popular infill strategy in European cities.

What is required to label a project as new urbanist? Does it need mixed use, traditional design, high density, an affordable mix of housing, pedestrian orientation, public transportation, or integrated street networks? While new urbanism approaches may advocate those elements, we rarely find European projects that include all. A wide range of projects may be listed as new urbanist: they appear on the CNU web site, for instance, or on a project list by new urbanist firms. Most of the ones in Europe, however, have only selected features.

For instance, the Borneo Sporenburg project in Amsterdam, in the Eastern Docklands, features high density (100 units per hectare) housing (ArchNewsNow 2002; Russell 2001). Descriptions of the project do not mention other uses in the project, and make it appear that the modern-styled housing may be fairly expensive. The project is described as an example of new urbanism essentially because of its density and infill location.

Karow Nord, Berlin, is another frequently-cited example of European new urbanism (see Box 5.1). As Bodenschatz (2003) indicates, though, in Berlin no one called it new urbanist. Instead, government saw Karow as a condensed suburb. The project involved 5200 subsidized modernist housing units in a high density village setting. The CNU web site lists the project as a form of urban infill (CNU 2004b), although other material describes it as a greenfield extension to the town of Karow (Gause 2002).

Among the CNU charter awards winners for 2001 (CNU 2003) we find Fonti di Matilde in San Bartolemeo, Italy. Some nine kilometres from the nearest city, this small greenfield site (less than 50 ha) will include 60 homes, a hotel, a spa, and a

Box 5.1 Karow Nord: rebuilding the German village

The rejoining of East and West Germany, coupled with the decision to reinstate Berlin as capital, set the basis for a housing boom in Berlin. The 1990s marked a new generation of housing development that provided the opportunity to implement fresh models for housing communities. Housing projects were concentrated in unused inner city areas and on undeveloped land at the urban fringe. In 1992, the Berlin Senate brought out a master plan for Berlin's home construction programme calling for between 80,000 and 100,000 subsidized homes. By 1995, new zoning plans allowed for 710,000 new housing units.

Karow Nord, a model project for Berlin's Senate, was one of many housing projects proposed to help fill the need for homes. Located in the north eastern part of Berlin, it was one of the most hotly debated of the new housing projects. The master plan was completed by Moore Ruble Yudell of Los Angeles. Construction started in 1994 to provide homes for approximately 10,000 residents.

Karow Nord presents many principles advocated by new urbanist-inspired planners:

Walkability and connectivity. A grid system of tree-lined streets, bicycle ways and quiet lanes connects parks, playgrounds, courtyards, schools, and daycares. All of the amenities, including library, banks, and shopping, are within a short walk.

Mixed-use and diversity. A small commercial centre provides all the needs for daily living: grocery, restaurants, stores, and professional services (see Figure 5.4). To a lesser extent, commercial spaces are also mixed throughout the community. Public facilities include daycares, schools, civic centre, library, teen centre, and municipal offices. Young families dominate the makeup of Karow; however, there is some mix of older and younger residents. The variety of housing types offers the opportunity for older people to stay in the community, and thus supports changing needs.

Figure 5.4 Karow Nord: modernist new urbanism
A mix of uses and a high density of apartment units make Karow Nord in Berlin an efficient use of urban infrastructure, but with contemporary building styles.

Mixed housing. The 'design code' allows four types of multifamily residential units based on the forms found in the traditional village of Karow. These include Housing Blocks, Karow Courtyards, Row Houses, and City Villas. The selection offers units supporting a range of lifestyles and price groups. Public funding was provided in three levels: one third classic social rental housing, one third subsidized rentals, and one third without funding (where homeowners pay market prices).

Quality architecture. The design code included guidelines for front yards, fencing, streets, facade details, roof lines, materials, and colours. The civic buildings are placed centrally and act as architectural anchors. The community proves that contemporary architecture can also be incorporated with design codes, where normally, 'copies' of traditional buildings dominate such schemes.

Traditional neighbourhood structure. The newly planned community is built into the structure of the surrounding traditional community of Karow. The edge of the new development reflects and coincides with the existing traditional structure. The housing becomes denser towards the centre where large public squares and parks are found. The district centre, situated in the neighbourhood core, offers both civic and commercial functions.

Increased density. Karow is much denser than most North American planned communities, with 111 units per hectare gross density (193 net). The density makes it possible to support public transportation and civic facilities like daycares, schools, libraries, and commercial uses.

Smart transportation. The city's light rail and bus system connects residents to the city centre within 30 minutes. Bikeways along main streets and quiet lanes support bicycle riders and roller-bladers, and connect open spaces and public facilities.

Quality of life. Karow Nord is attractive to people searching for affordable housing in a green setting within a short distance from the city. The public facilities and quality public spaces – including parks, squares, and playgrounds – are better than those offered in many newly planned communities.

Karow Nord meets many of the goals new urbanist planners set out to achieve. However, despite planning efforts, the community does not overcome some of the problems associated with conventional large-scale residential developments. The community was built on farmland surrounding the village of Karow. Although Karow Nord has many public facilities in place, the addition of almost 10,000 new residents has overloaded the infrastructure of the small village.

Despite a good public transportation system connecting Karow to the city centre, the community remains auto-dependent. Most residents must commute to the city centre to work. While the community has many shops, the big-box retailers attract shoppers looking for better prices and wide selection. Although abundant parks, playgrounds, squares, lanes and bikeways try to promote vibrant life on the street, the community remains lifeless. The open spaces and bikeways are not intensively used by residents.

The housing and social mix does not meet the levels planned. The private housing market has not been successful: only 169 of the 750 planned units were realized. In the effort to build more affordable housing to meet consumer needs, developers deviated from the design codes. Side-yard gardens were replaced with car ports; public lanes and parks normally found between houses were eliminated to allow bigger private lots. The large amount of subsidized housing without the homeowner contingent leaves a community that is not well mixed socially.

Some obstacles and problems were not foreseen in the planning process. Karow was built in two phases. The first phase contained all of the new urbanist features outlined above. Unfortunately, many elements of the design codes were not followed in the second phase for reasons including funding cuts, lack of commitment or coordination from government agencies, and unexpected market trends. During the late 1990s, economic conditions changed drastically. Many of the funding possibilities for private investors were cancelled. Developers saved money by cutting out the 'extras' required by the design code. For example, they reduced the size of front yards which had had a direct impact on the ability to rent first floor apartments on busy streets. The comfortable human scale of the first phase sometimes got lost in the second. As well, the open spaces did not live up to expectations generated in the early phase.

Planned communities with design codes require a greater commitment from municipal and provincial offices. Coordination between the planning offices is important to implementing the design codes. During the second phase the manpower was not available to coordinate the many designers and contractors.

Karow Nord boasts a wide range of well-designed open spaces. Unfortunately, maintenance costs are high. After completing Karow, the municipal government reduced the budgets for its parks department. In the early years, the open spaces were not well cared for and fell into a bad condition. Some committed community groups became involved in maintaining the public open spaces with success.

In the late 1990s, the City's direction for housing concentrated primarily on inner-city development to accommodate lifestyles which reflect the aging of society, smaller families, and more singles. Many new planning projects in German cities work on redeveloping the existing urban fabric. Through the process of de-industrialization, former railway yards, industrial areas and military land are becoming available within cities. These projects emphasize expanding the traditional city.

In the fall of 2003, the Council for European Urbanism presented a Charter outlining twelve challenges confronting European cities. Many of the points address concerns outlined by new urbanism supporters. Although there are many differences between North American and European cities, we see that sprawl, car-dominated travel patterns, and social segregation are common problems. The CEU adapts the basic new urbanism principles for European circumstances. As this group evolves, it will be interesting to see if the European supporters will create as much debate as did their North American counterparts.

Source: Trevor Sears

church. While the project features a mix of uses, it is a resort, not an urban setting. It will not be affordable; its residents and visitors will remain car-dependent. In this case, its urban form and design elements earn it the 'new urbanist' label as a 'neighbourhood plan' on the CNU list.

The English government has evidently become a strong advocate of urban villages: mixed use, high density centres for growth (ODPM 1999). The first project to call itself an 'urban village' in the UK was Bordesley in Birmingham in the early 1990s (Franklin 2003). The new development included a mix of housing and ownership types, but provided poor public transportation access. Franklin suggests that, in the case of Bordesley, proponents of the regeneration scheme found the title useful but did not see the need for an ideological commitment to all the associated principles.

The best known European example of development linked with the new urbanist movement is Poundbury in Dorset (see Box 5.2). The Prince of Wales commissioned Krier to do the site plan and many of the buildings in Poundbury. Just as Seaside and Celebration have become famous (or perhaps infamous) in America, so is Poundbury receiving an incredible amount of press and public attention. Krier's plan for a community of about 5000 is planned as four urban quarters with a mix of uses, and a village centre with retail (Thompson-Fawcett 1998). Prince Charles is quoted as having said in 1999 that 'Poundbury has merely tried to revisit those timeless principles that are best able to create a real sense of community' (cited in Gause 2002:162). At 39 units per hectare (Gause 2002), the project is denser than conventional suburban growth in the UK. It will include about 20 per cent social housing, and some light industry. Gause (2002) notes that Poundbury has less retail than originally planned (see Figure 5.5), and that units there fetch a premium of 5 per cent to 10 per cent over neighbouring properties. While its critics may find it inauthentic, in general people describe Poundbury as attractive (Hardy 2003).

Greenwich Millennium Village in the East London docklands area provides a vivid contrast with Poundbury. Development by 2004 included lots of apartment buildings and modern townhouses. Colourful, but definitely non-traditional, architecture by the Swedish architect, Ralph Erskine, sets the project apart. Some liken the look to Lego blocks piled high. The tall apartment buildings are bright and attractive, with beautiful landscaping and expansive views of river and wetlands (see Figure 5.6). An artificial lake and large ecology park draw attention to environmental issues (Greenwich Millennium Village 2004). To promote sustainability, the housing has many energy-efficient features. The ground-oriented town-houses away from the river views seem less attractive: they look like colourful boxes marooned in a concrete sea. By July 2004, the community already had a school, health clinic, convenience store in a temporary building, and a pub nearby; more shopping is planned in later phases.

Figure 5.5 Poundbury: classical market
With its pudgy columns and steep roof, Poundbury's village market makes a unique architectural statement. Instead of fronting on a regular street or a town square, it faces a large parking lot. (photo: Marc Ouellet)

Figure 5.6 Greenwich Millennium Village: Legoland apartments
Brightly coloured apartment buildings overlook the ecology park in this brownfield redevelopment project in the London Docklands.

Box 5.2 Poundbury: not quite real

On 10 January 2004, I departed London's Waterloo Station en route to Dorchester South Station. I had one mission, to see Poundbury, a neo-traditional development I had first encountered on the internet a few months earlier. On my arrival, I was grateful to find a row of taxis waiting, as I was unfamiliar with Dorchester. I got in the cab and obtained my first real perspective of Poundbury.

The cabbie provided a wealth of information, talking a few minutes even after I had paid my fare. He claimed that he and other local citizens really liked the development, even though it was a 'nightmare' for cab drivers: the streets are narrow and parking space is limited. He also told me that Poundbury has been responsible for an increase in the number of tourists in the area.

One thing that quickly caught my attention as soon as I left the taxi was how few people were walking on the footpaths. Cars zipped by as I waited at a pedestrian crossing. I must have stood a full minute before the steady stream of cars gave me a chance to cross.

As I made my way through the streets and footpaths leading into the inner courtyards set aside for parking, I examined the architecture. The buildings reflect traditional and classical styles. Materials in the exterior cladding include brick, stone, and stucco. Lot coverage is high and building heights are mostly two to three storeys. While building scales are similar, giving coherence and harmony, variety is achieved through architectural forms and features, and alternating building materials.

Poundbury had an interesting mix which besides homes included a chocolate factory, veterinary clinic, cancer care facility, nursing home, information technology company, small business development group, publisher, and cereal factory. However, many basic services were missing.

The designers have tried to alleviate the automobile's footprint in Poundbury. Inner courtyards surrounded by buildings and high walls serve as internal parking areas. Garages are hidden in the courtyards and in alleys. Measures are used to obstruct the view of garage doors to passersby. The inner-courtyard parking areas are accessed via footpaths and porte-cochères which increase permeability. However, the high walls and lack of windows fronting on some of the paths did little to reassure me that these would feel safe at sundown.

Needless to say, I saw many more people driving than walking. The town square serves as a parking lot rather than a public space. Some of the newer phases don't seem to hide the automobile as much as did the original phases. Garage doors front directly on streets, and some of the newer townhouse developments have front yard parking. Many cars park across sidewalks. I saw an elderly woman pushing a walker on wheels in the middle of the street because cars obstructed the sidewalk; the fine gravel used for the footpaths was quite unsuitable for her use.

Later I watched helplessly as an elderly man slipped on the gravel and fell in the square. The gentleman conversed with me for a few minutes, answering my questions and volunteering bits of information. I asked him if the square was originally intended for parking. Not really answering my question, he claimed that it was closed off to parking a few times during the summer for an outdoor market.

A saleswoman in the developer's office told me that builders submit their own plans which must respect Krier's guidelines and meet the requirements of the architect assigned to oversee compliance. She indicated that Poundbury was the best market in the area. The selling prices of the dwelling units were higher than for Dorchester.

My final conversation within Poundbury was with a couple in their mid-thirties. They lived in Dorchester proper. I met them as they left *The Octagon*, the local coffee shop. They had different perceptions than the other people I had spoken to that day. When I approached the man, I asked him if I was standing in the town centre of Poundbury. He responded, 'yeah, not much of a town centre is it?' They said that Dorchester locals didn't think much of Poundbury. The woman insisted that Poundbury may look nice, but it didn't fit with the local architecture and the modern reality. The couple claimed that residents of Dorchester resented Poundbury, since the people moving into Poundbury were not from the local area and tended to be cliquish. As they left me, the man volunteered a final remark: 'Well, they have a nice coffee shop.'

Then I went into the market. It looked interesting from the exterior, but the interior seemed quite ordinary. In essence, the town was very much like the market. On the surface it reflected an aesthetic ideal, but when one looked beyond the surface, into the inner workings of social processes and human behaviour, the result was flawed. An enduring dependence on the automobile, lack of self-sufficiency in services, and limited socio-economic diversity, contradict the concepts of a healthy neighbourhood.

After walking around Poundbury for two-and-a-half hours, I decided to head to the train station on foot. I must not have paid attention to the route taken by the taxi, since I quickly became lost and ended up walking around Dorchester for about an hour before I located the station.

I was pleasantly surprised with what I saw in Dorchester and couldn't help comparing it with Poundbury. Dorchester had a more varied building form and a real downtown that was quite vibrant for the size of the town. Dorchester had public amenities, such as parks and schools, missing in Poundbury. The socio-economic diversity in Dorchester seemed greater. But what I found most apparent when I walked around Dorchester were the many pedestrians and children. I left for London thinking that Dorchester felt closer to the ideal of Poundbury than what the new development had achieved. At least Dorchester felt real.

Source: Mark Ouellet

The site is well linked to public transportation networks with frequent buses to the nearby tube station. Although housing in the project is not inexpensive, the project has many features seen as important in new urbanism.

In recent years, American new urbanist firms like DPZ have won commissions in Europe, raising the profile of new urbanism. As Bodenschatz (2003) notes, Europe has had neo-traditional architects for some time, so that element is not new; a strong urbanist tradition characterizes the work of architect-planners like Rob Krier. Bodenschatz sees new urbanism's commitment to managing sprawl as its most useful contribution to European urbanism. While the American New Urbanism preoccupation with street pattern does not translate well in the European context, for instance, the attention that new urbanism pays to regional planning resonates with European concerns about sustainability. We might be tempted to conclude that the principal features of the European new urban approaches are more closely aligned with ideas of sustainable development and healthy communities than they are with the design prescriptions of American New Urbanism. European new urbanism has aside some of the principles that North Americans have associated with the movement.

This brief review only highlighted a few of the many projects in some ways linked – or claimed to be linked – to new urbanism, urban villages, and the urban renaissance in Europe. Clearly, though, new urbanism is *not* the dominant ideology in contemporary development in Europe. What are the conventional European patterns? High land costs and a tradition of relatively dense urban and suburban living generally result in smaller lots and homes than we see in conventional development in North America. The suburbs continue to grow around many cities as affluent households look for comfortable neighbourhoods in which to raise their families. Efforts to ensure more compact form and to reduce car dependency run into several fiscal and cultural barriers.

Challenges and barriers to meeting the targets

Even with widespread agreement about the need to manage traffic, increase densities, and control sprawl, meeting the targets and employing the strategies suggested by new urbanism is not proving easy. In this section, we briefly review some of the barriers to changing development practices in the ways that new urbanism approaches suggest.

As is the case in North America, the practices of investment bankers and the rules that govern planning may limit opportunities to advance the principles. Stead and Hoppenbrouwer (2004) say that investment practices and zoning rules restrict mixed use developments in the Netherlands. Projects such as the Rue de Laeken in Brussels and Crown Street in Glasgow have found the leasing of commercial properties slower than anticipated, delaying the achievement of a robust mix and undermining the profitability of projects. Where government is a major partner in projects, or where commercial components of the project may be part of later phases, then lenders may be more willing to take risks. Otherwise, conventional developments may seem safer for investors.

In the UK, governments are promoting urban villages as an option: projects vary in size from one to 300 ha, from 160 to 15,000 people. Urban villages seek to combine urban features (like ready access to public transportation) with village features (like daily needs ready at hand in a walkable environment). They typically have a limited range of mixed use, with some shopping. They should present opportunities for local employment. Biddulph *et al.* (2003) conducted a study that identified 55 urban villages: not a comprehensive list, but sufficient to identify key trends. They suggest that practice reveals a gap between the ideal and the real in the urban village movement in the UK. The apparent consensus about what they should achieve is illusory. Franklin and Tait (2002) argue that the urban village is more an artifact of the rhetoric of planning discourse than a real urban form on the ground. The term is used to give legitimacy to projects and decisions, but does not signify a major change in development patterns.

Demographic and cultural factors make it hard to achieve the compact city. Even though the density of units is increasing, declining household size means more dwelling units are needed for fewer people. De Roo (2004) says that the Netherlands expects to need to add one million housing units despite a stable population, simply to accommodate the trend to smaller households. Relocation of populations from declining areas (like the industrial north of England or the former East Germany) is adding to growth pressures in expanding cities.

Although urban living is rapidly becoming the dominant choice in the world, Stead and Hoppenbrouwer (2004) say that people in the UK and the Netherlands prefer rural living. They want air quality, safety, quiet, and privacy which they may see as lacking in the city centre. Many urban workers have moved far away from their places of work, willing to commute long distances to homes in a village or in a suburb (BBC News 2003). Europeans are increasingly suburban residents (Angotti 2002). Attracting people to return to the city centre can prove a challenge.

Strategies that push for higher densities may generate reactions from local residents. For instance, a neighbour of the proposed Upton Village project in Northampton told me in July 2004 that he worried that the government would 'cram many little boxes' into the area. Recently Poundbury residents protested an apartment building they feared would drive up density in the village (Langdon 2004d). In the Netherlands, distress over town cramming has led government to talk more about the 'comprehensive city' than the 'compact city' (Stead and Hoppenbrouwer 2004). In the UK, the focus is on 'sustainable' cities and an urban renaissance. Whatever the terms used, convincing suburban residents of the need for greater density is a hard sell.

Is the effort to enforce compact form and infill development driving up housing costs? Developers in the UK think so: Wimpey Builders have complained that planning guidance encouraging redevelopment on urban sites will restrict the housing supply in the South East (Aldrick 2004). Tight supply certainly contributes to high costs in many parts of the UK, so it seems reasonable to think that compact form should help ease the situation.

As has happened in the USA, Europe is experiencing the worrying trend of gated developments (Angotti 2002; Webster et al. 2002). Gated enclaves allow residents to close themselves off from the city. Even areas within public housing projects have erected gates in the UK (Manzi and Smith-Bowers 2003; University of Westminster 2004). Security has become a big issue, especially in Britain. Video surveillance cameras, and signs alerting passersby to the security cameras, are omnipresent. A vigorous debate has erupted between the proponents of new urbanism and the advocates of crime prevention through environmental design in the UK (Knowles 2003; Steuteville 2003b; Town 2003). Knowles (2003) argues that new urbanism contributes to crime through higher density and mixing, and by encouraging street layouts that facilitate easy access for criminals.[2] The *Safer Places* guide produced by the British government (ODPM 2004b) suggests that cul-de-sacs provide safer streets. Target hardening has also become a popular strategy for promoting safety. While the notion of 'eyes on the street' has gained considerable currency in North America, Europeans generally keep their curtains or shutters closed for privacy. Walls, gates, and enclosed courtyards seem more popular than the open city model associated with new urbanism.

Where heavy traffic is such an issue in the narrow streets of European cities, open street networks may present a risk. While North American New Urbanism promotes interconnected streets to distribute load, in Europe one more commonly sees bollards used to close streets, and speed bumps, chicanes, and roundabouts to slow traffic. The open grid of the American city, so popular with North American New Urbanists, does not translate well to Europe. European new urbanism shows greater concern about connectivity for pedestrians and cyclists while carefully controlling cars in residential environments.

The challenge of how to achieve and maintain mix is as much an issue in Europe as it is in North America. Local retail convenience stores are closing as malls and big-box retailers siphon away custom. As car ownership increases, people drive greater distances to shop. Growing affluence has provided households with access to larger refrigerators and freezers that have the potential to transform shopping behaviour. A viable mix of uses may be slow to arrive and hard to keep when consumer shopping behaviour is changing so dramatically. For instance, Tait (2003) found that residents of West Silvertown urban village in the London Docklands did most of their shopping by car despite efforts to create mixed use locally. Mixed use remains important to the least affluent in the community: Tait notes that those in social housing (30 per cent of housing there) were more likely to not have a car, to want local grocery stores, to work locally, and to use village hall facilities. Some 60 per cent of Poundbury residents surveyed by Thompson-Fawcett (2003a) were dissatisfied with the shopping facilities. Keeping retailers viable in residential neighbourhoods proves one of the most difficult of the challenges of new urbanism approaches everywhere.

As Tait (2003) notes in West Silvertown, and Thompson-Fawcett (2000, 2003a) reports for Poundbury, mixing housing types in new urban projects does not ensure that the residents of the housing interact with each other in a positive way. Those in social housing tend to socialize with others in social housing, not with those in private housing. The polarization of the classes and household types continues in new developments, despite the social aims of the projects. Thompson-Fawcett (2000) describes tensions between households in different kinds of units (public housing *v.* private homes) over children's activities in Poundbury: conflict resulted in the posting of signs to restrict ball play in lanes. Good design cannot overcome all the social barriers to integration. Thompson-Fawcett (2000:287) argues that the strong social agenda of the British urban village movement 'upholds the "traditional" nuclear family; neighbourliness and decency; community commitment, responsibility and self-policing; and orderly social relations'. It provides, she says, a managerial way of dealing with the urban crisis. Given the continued polarization in the new projects, however, it seems unlikely to resolve the underlying inequities in our cities. Moreover, Biddulph (2003b) notes that while professionals may romanticize the traditional village, for locals the village may represent a lack of choice and mobility that they hope to escape. Class differences cannot be wished away by good design.

Are the new urban approaches reinforcing vernacular architecture, or creating a new kind of uniformity? Some of the projects have made admirable efforts to build according to local patterns. For instance, the Rue de Laeken project fits well in the context of its district in Brussels. In some cases, though, it appears that designers have fixed on particular design principles that they apply regardless of the circumstances. A case in point: the columns on Leon Krier's buildings in Poundbury look suspiciously like the columns Krier and Gabriele Tagliaventi used in Alessandria, Italy (see Steil with Salingaros 2004). Krier's original plan for Poundbury was said to have resembled a Tuscan village, not rural Dorset. When a limited number of designers and architects adhering to a template of architectural principles for inspiration enjoy such a high profile in a movement, we may see a degree of internationalization, especially in the buildings for commercial and public use. While the same held true in classical times (as the Romans employed a single set of design principles from north Africa to the Scottish border), a movement that prides itself on sensitivity to the local vernacular must find the standardization frustrating.

The language of sustainability appears more commonly in European discussions of new urban approaches than it does in the American context. Does this suggest that Europeans are making greater progress in dealing with environmental degradation? Europe faces some significant environmental problems: poor quality of waterways, loss of forested land, disappearance of ecosystems like hedgerows and wetlands. What kinds of urban development are 'sustainable' in that context? Difficult to say. The ecology park at Greenwich Millennium Village in London makes a nod in the right direction. It provides a refuge for wetlands species in a region significantly modified by urban and industrial development through the centuries. A block or two away from the ecology park, though, vast areas of pavement unnecessarily disrupt landscape function by reducing permeability. Many of the new urban projects could do much more to restore the environment: plant more trees, use energy-efficient building materials, erect rooftop gardens, and provide allotments for growing food. One of the key benefits of the early garden city movement was its commitment to gardens and gardening; that interest in providing spaces for vegetable and flower gardens reflected a concern about the local residential environment and its quality and healthfulness. With the current urban focus on getting people mobile – to work, school, shops – designers treat the landscape around the home as a transition area, rather than as an activity zone. The ambit for judging what contributes to sustainable options seems quite narrow.

European development already occurs at higher densities than we see in North America. The new suburbs, even in conventional forms, show the same pattern: small lots, small homes, and a high proportion of attached or apartment units. Suburbs in Europe often have reasonable access to public transportation and a greater focus on energy efficiency than suburbs in North America. That said, few find that the endless new building in Europe qualifies as sustainable. The search for alternative paradigms for development has led governments to consider options like urban villages. But, as Thompson-Fawcett (2000:278) says: 'The urban village is posited more in the form of a cloudy paradigm than a prescriptive model, making it especially vulnerable to liberal interpretation.' The urban village may not guarantee sustainable communities.

Too often seen as an American idea, new urbanism has been slow to gain adherents even as it has influenced the discourse about the good city in planning circles everywhere. The label of new urbanism is certainly less commonly used in Europe than in North America. Often we find it applied retroactively and at a distance to European projects, as if the advocates of North American New Urbanism wish to suggest that their movement has a greater international dimension than it does. That situation may, however, be changing. An article in the *Guardian* in July 2004 (Campbell 2004) reported that the president of the Royal Institute of British Architects called on his colleagues to support new urbanism. Subsequent letters to the editor of the paper reflected a degree of resistance from British urbanists. Malcolm Smith (2004:online) wrote that starting 'from the sterile strategy of "new urbanism" risks repeating past mistakes by importing inappropriate theories'. Richard Rogers, chair of the Urban Task Force that proposed the urban renaissance in Britain, affirmed the principles of mixed use and a compact multi-nodal city while challenging the claims of new urbanism.

> But these principles are not the sole preserve of 'new urbanism'. There is little new in this movement except for its blending of well-established urban design principles with a romantic neoclassical style that often tumbles into tawdry pastiche. This is no leap forward in the 21st century.
>
> (Rogers 2004:online)

The European debate, begun with criticisms of the modernist approach in presentations to RIBA by Leon Krier and Prince Charles in the 1980s, is now fully engaged. Will the urban renaissance and urban villages movements find themselves fully fused within new urbanism, or will RIBA's conversion prove short-lived? Will Europeans continue to chart their own approach to urban design, or will we see a growing international consensus on the strategies necessary to plan the good community? It may be too early to say, but it seems reasonably safe to predict that diversity of approaches and solutions will likely remain the outcome in European urbanism.

CHAPTER 6

MODERNIZING URBANISM: NEW URBAN ASIA

The context of East Asian urbanism

Given the diversity of East Asian cities, I take a great risk in trying to make any generalizations about the patterns of new urban approaches in the region. Hence I will narrow the focus to what I know best: Japan. In this chapter, I begin with a brief overview of traditional and contemporary urbanization trends in East Asia, before considering recent approaches to urbanization in Japan in greater detail. Japan is an economic powerhouse with an urban tradition that dates to the eighth century AD. Its people enjoy a standard of living comparable to that in the West, while its cities exhibit some of the features that Western authorities define as sustainable: high density, compact form, mixed use, and extensive use of public transportation. A brief review of Japan's approach to urbanization casts some doubts on the universality of new urbanism premises and promises.

Urbanism in East Asia proceeds by some of the assumptions implicit in new urbanist principles: compact high density cities; integrated mixed use; reliance on transit-oriented networks. At the same time though, East Asian urbanism pays little attention to the public realm, has relatively little interest in heritage conservation, increasingly accommodates the private automobile, and rarely favours classical design principles or vernacular traditions. In many ways, East Asian new urban approaches favour modernist principles in design and urban form. With strong traditions of crowded cities with narrow streets and declining environmental quality, these cities define their problems differently than do North American and European nations, and look for different solutions. Their practice reveals that rather than guaranteeing sustainable and healthy cities, high density and mixed use can generate problems that undermine urban quality of life and motivate residents to demand changes. New urbanism, with its principles designed to respond to the issues generated by sprawl, may have little to offer dense cities.

A legacy of strong population growth sustained by bountiful food production has generated high density through much of East Asia. As Heikkila and Griffin (1995) note, Asian cities are diverse at the micro level, but homogeneous across the wider urban level. Homes, workshops, small factories, gardens, and stores abut each other in districts throughout cities. Shelton (1999) says that to Western eyes the jumble of uses may seem like chaos, but this historic character of Asian cities contributes to their dynamic vitality and economic strength.

Confucian tradition has had a major impact on urban form in East Asia. Heikkila and Griffin (1995:270) argue that urban form may be seen 'as social discourse that articulates the substance of culture and discloses a society's underlying structures of meaning'. Urban form reveals cultural values and meanings. Adopted as state ideology 2000 years ago in China, Confucianism spread to influence Korea and Japan. The philosophy promotes hierarchy, conformity, and service. It had an immense influence not only on the structure of government and society, but even on the patterns of urban development in East Asian cities.

Asian cities have experienced rapid growth in recent decades. Marcotullio (2004:39) calls the development 'compressed and telescoped'; in other words, in a relatively short time period, East Asian cities progressed through stages that in Europe and North America happened over decades or centuries. The processes of globalization are compressing time and space so that cities now modernizing may jump from 'Third World' to 'post-modern' conditions, or reveal both simultaneously (Harvey 1989). Such growth certainly disproves evolutionary theories of urban development. More importantly for practice, though, Marcotullio (2004) notes that it means that cities can simultaneously experience both the basic sanitation problems endemic in the cities of the developing world (like poor water quality and inadequate sewerage) and the contemporary dilemmas of the post-industrial city (such as persistent toxins, heavy metals, and rising carbon dioxide emissions). They face a double challenge.

A legacy of compact living and population growth makes Asian cities among the most densely populated in the world. Kwang-Joon Kim (2004) reports that in 2000 the population density in Seoul was 16,356 per sq km; in Tokyo 13,092 per sq km, and in New York 9719 per sq km. As Grant et al. (2004) note, though, problems associated with density (such as pollution, traffic, noise, lack of privacy, lack of open space) undermine the quality of life in Asian cities. While Western models of the vibrant and sustainable city inevitably point to higher densities and compact form as essential elements of the good city, the Asian experience reminds us that density is not a universal solution to the issues of the contemporary city. Relentless growth undermines efforts to manage the pressing problems of urban life: instead of being able to focus resources on improving conditions in older parts of the urban fabric, governments face competing claims for investment. Moreover, Asian cities reinforce the conclusion that rising affluence drives urban sprawl: as household incomes rise, car ownership is growing throughout Asia, and people look farther to the peripheries for opportunities to buy affordable homes.

The solutions Western planners offer for urban dilemmas may not work in Asia because the context is so different (Marcotullio 2004). East Asian cities are already compact, with narrow streets and good public transportation. Those features have not solved all the urban problems, but instead generated different ones worth noting.

The 'compact city' model or new urbanism may not help East Asian cities with their urban problems. The Asian experience suggests that the good community may be compact, but it also needs to be clean, green, and participatory.

I begin with brief overviews of urban issues in China and South Korea to set a regional context for developments in Japan, before proceeding to a more extensive description of Japanese urbanism.

CHINESE URBANISM

China has the longest history of urbanism in East Asia. Its earliest cities date to the second millennium BC. To ensure that we should not think of traffic problems as restricted to the post-automobile age, Wu (1986) reports that the city of Linzi experienced significant traffic jams in the first millennium BC. Cities that became centres of administration and trade for expanding empires attracted throngs of visitors. Traffic problems may thus be intrinsic to successful urban development.

The Chinese developed a distinctive model of urban form by the first millennium AD. The grid layout of the capital city had roads oriented to the cardinal directions. The palace lay to the north of the city and looked south. The overall plan represented the hierarchy and order implicit in Confucian thinking (Wu 1986). While housing and work places would have mixed in the various urban areas, the prescribed layout determined function, with trades and groups designated particular districts within the grid.

The Tang dynasty rebuilt Chang'an (Xian) as its capital in the seventh century. As many as one million people lived within the wall, and perhaps another million in the suburbs outside (A. Wright 1967). The layout followed the hierarchical pattern so popular for Chinese capitals. Courtyard houses faced south, while walled and gated districts were inhabited by specific trades (and closed off at nightfall). The city was compact with a mix of uses throughout. Grand streets separated walled districts and led from the southern gate of the city to the administrative district in the north. In cities like Chang'an, the grid represented the spatial manifestation of the social and political hierarchy of the society (Grant 2001a): the more powerful the family, the farther north they lived in the city. Capital cities relocated frequently in Chinese history, as regional resources and dynastic changes made new sites more attractive. While urban form changed slightly through the centuries, the basic layout of the capital cities tended to retain the patterns developed at Chang'an (Wu 1986). Visitors, traders, and diplomats from throughout East Asia visited Chang'an to return home inspired by its glories.

The great city of Hangzhou (Hangchow) became the Sung Dynasty capital in 1138. Its rulers built hospitals, parks, gardens, beautiful buildings and grand bridges over rivers and canals (Girouard 1985). When Marco Polo visited in 1274, the grandeur of public places and private palaces astounded him. Telling his tales back in

Venice, Polo inspired several generations to travel to the Orient to see the urban developments for themselves, and ultimately to emulate them in Europe. The Chinese city thus became a model of good urban form for developing European polities.

When the Communists came to power in China in 1949, they had little respect for the city. They associated the elegant touches, formal layouts, and courtyard housing of the cities with discredited régimes. Bonavia (1978) says that the anti-urban policy implemented by the new government led to the loss of a considerable amount of urban and natural heritage. The government certainly tore down old districts to replace them with modern apartment buildings and factories. While the Chinese have a saying that 'all waste is treasure' (Bonavia 1978:40), they may not view their own heritage as worth preserving.

The Communist revolution brought rapid modernization and industrialization to China. The process of economic reforms that began in 1979 initiated another spurt of growth. The gross domestic product of China has expanded at 8 per cent a year since the early 1980s (Auffhammer 2004). The focus on economic growth above all else has contributed to significant environmental problems, such as poor air and water quality. Over-population has long presented an issue for China, and an infrastructure deficit undermines quality of life in some urban centres. At the same time, a new middle class is building monster homes in the suburbs, contributing to new forms of sprawl (Cohn 2004).

To encourage an indigenous automobile industry, the Chinese government has targeted growth in car ownership as an economic opportunity over the next decade. As of 1995, China had eight cars per 1000 people, while India had five, the UK 406, Japan 538, Canada 594, and the USA 750 (MTC 2004). In recent years, though, car ownership has been growing rapidly in China (Auffhammer 2004; Handwerk 2004). Consequently, traffic jams, high rates of auto deaths, and air pollution are worsening.

Beijing has won the right to host the 2008 summer Olympics. To prepare for the event, authorities are cleaning up the city. In some cases, this has meant clearing out older areas. Geoffrey York (2004) describes the hopeless desperation of the residents of old courtyard housing off alleys, known as *hutong*. This traditional form of housing, with some homes more than a century old, is disappearing, even though buildings may be designated as 'heritage' properties. Traditional design principles seem to hold little interest for the Chinese: modernist design is commonplace. In fact, Barber (2004:R1) writes, China exhibits 'free wheeling architecture on a scale unseen since the Bauhaus'. Most new development occurs in the form of medium and high-rise apartment blocks, often within gated communities (Giroir 2003). China seems to have little interest in the principles associated with new urbanism in North America or Europe. Its new urban approach is resolutely modern: in its rush to modernize, it pays limited attention to issues that Western designers associate with good urban form.

SOUTH KOREAN URBANISM

Like other East Asian nations, Korea has a long urban tradition. Pyongyang is one of the oldest cities in the world, with roots in the third millennium BC (Tae 1962). Later Korean capitals show the influence of Confucius and the Chinese city of Chang'an, with the hierarchical grid layout that came to dominate capital cities through the region.

Now a city of 20 million people, Seoul had 46.2 per cent of South Korean's population in 2000 (Lee 2004). Areas some 30 to 40 km from the centre of the city are in the process of becoming suburbs. Lee (2004) suggests that land use regulations in fringe areas cannot accommodate such phenomenal expansion while maintaining quality of life. Because the Housing Ministry plans housing in new towns in Korea, the districts themselves lack the power and resources to implement the necessary range of services and adequate access to public transportation to meet new demands. Managing urban growth effectively requires more extensive coordination and integration of town planning (Lee 2004).

As Koreans become more affluent, they want more living space. Traditionally, the average per capita dwelling size in Korea has been well below that of other economic powerhouses: for instance, in 1998, Koreans had 17.3 sq m per person, while the Japanese had 29.4, the British 30.4, and the Americans 55.8 (Lee 2004). This desire to see improvements in the standard of living drives suburbanization as households move farther from the city to find space to meet their aspirations.

Kim (2004) notes that government policy created apartment district overlay zones in Seoul that allowed 2.75 to 3.0 times lot coverage. These intensification provisions led to processes of land redevelopment. In some cases, government has consolidated land into larger parcels and removed older housing for redevelopment at higher densities. In other cases, private landowners put their parcels together to form a redevelopment group. Usually redevelopment projects replace low-rise houses and walk-up apartment buildings with high-rise towers. The rapid growth of such projects results in larger apartments, but also in a loss of affordable smaller units. Squatters and poorer households are forced farther from the city centre. Kim (2004) reports that the government now requires that 20 per cent of new units occupy fewer than 60 sq m to try to address the need for affordable units. The spate of redevelopment has raised fears that densities may be too high already in Seoul, and that new urban policies should reduce those densities to improve the quality of life for urban residents. Government has been pressured to down- zone some areas to limit the demolition and redevelopment of older style low-rise residential districts.

The situation in Seoul highlights the dilemma of using intensification as a solution to the problems of housing supply and affordability. The city already has a tight urban form (Kim 2004). Many experts worry about the possibility of long-term decay in high-rise apartment districts. Increasing densities in Seoul clearly affect air quality by concentrating pollutants. Intensification contributes to the loss of the green belt around Seoul, the cutting of mature trees, the disappearance of historic building patterns, and potentially dangerous construction on hillsides. The mass transportation system, already well-used, may lack the capacity for increased volumes. Kim (2004) asks whether the city really can handle more density without reducing the quality of life of urban residents. While increasing densities offer a means to some ends, as an end in itself density cannot guarantee a good community.

NEW URBAN APPROACHES IN JAPAN

Our brief review of trends in China and Korea shows that patterns often associated with modernism continue to influence development in East Asia. High-rise towers, class-based segregation, car-oriented suburbs, and land use patterns that feature less fine-grained mixing than traditional cities exhibited are appearing with greater frequency. We find no indications that these booming regions find new urbanism attractive. In this section we look at new urban approaches in Japan to see if new urbanism speaks to the problems that face Japanese cities.

Japan urbanized later than China and Korea, but was influenced by the same urban philosophy. The first permanent city appeared in Japan in the eighth century. Earlier court capitals moved with the ascension of new emperors. In AD 710 the court established a capital at Heijokyo (now known as Nara). For political reasons, in 794 the capital moved to Heiankyo (Kyoto), where it remained for 1100 years. The layout of early Japanese capitals followed the Chinese tradition, and reflected the exchange of people and information with the Chinese court at Chang'an. A regular grid oriented to the cardinal directions had the palace in the north and a long processional avenue from the southern gate to the northern administrative district (Nara n.d.).

One of the earliest novels written, *The Tale of Genji*, by a court lady, Murasaki Shikibu in the tenth century, reveals the problem of traffic jams in Heiankyo (I. Morris 1964). Carriages constituted status symbols in the capital. Only the very highest ranks had the right to employ hand-drawn carriages. Others used open carts pulled by animals; while uncomfortable, these wagons provided an essential means of transportation for court officials and their families. On significant occasions, carts and hand-drawn carriages crowded the wide roadways, jostling for position.

Many of the broad avenues, from 25 to 50 metres wide, turned to mud with heavy rain. The streets had grand dimensions, but simple materials. Unlike the Chinese capital, which became a vast metropolis and trading centre, only part of the

Japanese capital became fully occupied: Suzaku Oji, the main north–south road, was 75 metres wide, yet it passed areas of the city abandoned to nature. The idea of a grand capital proved far ahead of its time. For the most part, Japan remained an agricultural nation with little demand for urban centres. Villages and towns developed as places of trade and pilgrimage, but no other large cities appeared for centuries. In the capital, simple and modest wooden palaces lay hidden behind dark walls and earthen banks. The homes of officials grouped around courtyards open to the south, in the Chinese fashion. Outside the capital, the country established a vibrant urban tradition only after the reunification of Japan under military rulers in the late sixteenth century. Appreciation for the rural way of life remained strong until modern times.

The Tokugawa Shogunate initiated a period of urbanization and expanded trade in the seventeenth century (J. Hall 1970). To hold territory, the government built castle towns according to a standard plan (Shelton 1999). The plans reflected the order and hierarchy of Confucian philosophy. The castle – like the palace before it – was usually located to the north of the site, or in the north–centre of the plan, with a grid of streets running north–south and east–west. Lot sizes and location reflected the social position of the inhabitants (Grant 2001a), with *samurai* enjoying large lots near the castle, and the artisan ranks grouped in districts on smaller lots to the south. Shop-keepers typically sold their wares at the street-front of their homes, and worked and lived behind the stores; poorer residents inhabited homes off alleys (Shelton 1999). With taxes calculated by the amount of street frontage, long narrow lots totally covered by low wooden buildings typically resulted (see Figure 6.1).

During the Edo era (seventeenth through mid-nineteenth centuries), Japanese

Figure 6.1   No 'eyes on the street'
In earlier times, merchants sold their wares in the fronts of narrow houses in Kyoto. With shutters and frosted glass facing the street, there are no 'eyes on the street' in the contemporary Japanese city.

cities had little open space except areas for military use within the castle keeps. The towns and cities were places for work, trade, and administration. The public infrastructure provided roads, bridges, canals, and castles. Unlike the rulers of European cities in the Renaissance, Japanese lords felt no need to invest in a beautiful and elaborate public realm. The moral code of the *samurai* promulgated an aesthetic that created lovely private gardens for contemplation and ritual, but limited spending on urban design (Nitobe1969)

Most of the buildings, except for the castles, were made of wood. If fire hit – and it proved a constant risk – residents could easily dismantle many buildings for removal to safety. People kept few possessions in the houses. Workers' districts had high densities, mixed use, and very little open space. Blank, dark walls or earthen embankments surrounded the homes of the more affluent.

Outside the original planned area, roads grew in a haphazard way in the spaces left between buildings (see Figure 6.2). In Edo (Tokyo), the centre of military administration, traditional Japanese urbanism resulted in narrow winding streets, low and dark wooden buildings with few windows on the street. By contrast, a cacophony of colourful signs and advertising banners illuminated commercial districts. The traditional city had few public spaces: shrines, temples, and bridges offered the only real

Figure 6.2 Winding alleys lead home
Narrow lane ways reflect a time before the automobile, when the streets or alleys were simply the spaces left over between houses. This street in an old part of Nagoya features a mix of uses.

public meeting places or areas of relaxation for ordinary people in their limited free time (Shelton 1999). The residential areas of the Japanese city tended to turn inward, away from the street.

In 1868 the shogunate ended, and the Meiji restoration of the authority of the monarchy led to a period of rapid modernization and Westernization that lasted through the 1920s. Many new public buildings took on Western styles. Some government buildings and banks featured the classical revival elements of columns, stone carvings, and symmetry popular in the West during the period. Architectural students from Japan travelled to Europe to study at the École des Beaux Arts, and to learn from notables like Le Corbusier. Famous modernist buildings, like Frank Lloyd Wright's Imperial Hotel built in Tokyo in 1915, provided vivid contrasts with traditional urban patterns.

During the 1930s, military governments in Japan rejected Western customs and reinforced traditional values. After the Second World War, however, the nation again embraced modernism, and really never turned back.[1] While many private homes followed Japanese architectural conventions with overhanging eaves and wooden walls, public buildings and commercial architecture tended to reflect the sometimes brutalist modern style of the era. Districts built during the post-war period reveal the eclectic tastes of Japanese urbanism; they typically include a jumble of building types, sizes, styles, and uses, and a clutter of signs, wires, and poles (Shelton 1999), as Figure 6.3 illustrates. In suburban areas, modern interpretations of traditional housing forms may appear alongside kit houses called 'American' or 'Canadian' style: a smorgasbord of design options adapted for the Japanese market.

Some 45 per cent of the Japanese population live in a narrow belt running from Tokyo, through Yokohama, Nagoya, Osaka, and Kobe: an area many call Tokaidopolis (Karan 1997). This heavy concentration of the national population in a small area is paradoxically counterpoised to simultaneous signs of de-concentration: people are moving to the suburbs of these cities (Sorensen 1999, 2002; Wegener 1994). Smaller cities (under 100,000 population) are shrinking as young people head to larger centres for education and work. Declining birth rates threaten the potential for long-term growth in the economy as well as in the cities. While Western nations look to immigration for population growth, Japan has not historically welcomed migrants in numbers necessary to maintain its population levels over the next century. We might expect population decline to have significant effects on urban networks over time.

Several challenges face traditional Japanese urban patterns of high density mixed use. High land prices make housing extremely expensive in Japan. The demand for space for business in central Tokyo is pushing out residents and hence reducing urban residential densities (Fujita and Hill 1997). The search for affordable housing, or more spacious housing at better prices, forces households to look farther from the city centres. Affluence gives people options, and growing rates of car ownership

Figure 6.3 Eclectic mix of styles in Tokyo
Contemporary Japanese cities have the most diverse architectural mix found anywhere. Traditional structures vie for attention with the most audacious innovative designs. The conventions of Western tastes do not apply.

facilitate suburban commuting and shopping patterns. The growth of suburban retail malls contributes to the decline of traditional shopping precincts, especially in smaller cities. In other words, many of the same pressures that led to reduced densities in Western cities threaten Japanese cities.

Planning and land policies have facilitated sprawl. The legal system in Japan gives property owners inalienable rights to build housing on their land, regardless of parcel size or available services. Agricultural subsidies and taxation policies encourage farmers to keep land in agricultural use even as it becomes surrounded by urban uses (Sorensen 2002). Land is highly valued in Japanese culture, and families prove reluctant to alienate it. Rapid escalation in land prices in recent decades reinforced the belief that farmers should not sell their land until it reaches its ultimate value. Hence, if farmers need money, they typically sell off small parcels. Developments under 1000 sq m could avoid the requirement for planning permission or the demand to provide urban infrastructure; this encouraged the development of small suburban developments (e.g., half-a-dozen homes) with no urban services such as piped water or sewerage (Sorensen 2002). Thus the periphery of the major urban centres exhibits a mix of uses: farming, industry, institutions, and scattered housing in close proximity.

These areas have few urban services and poor access to public transportation. The resulting pattern of sprawling mixed urban and rural uses is variously called scatteration (Hebbert 1986), interdigitation (Mather 1997), or land fragmentation (Sorensen 2002).

Japanese urban residents face extremely high costs of living because of the high densities in their urban areas. Recent surveys put Tokyo first and Osaka fourth among the most expensive cities in the world (CNN 2004). In the largest cities, dwelling units are consequently quite small. High-rise units are commonplace. Few people find housing near their places of work. The search for something affordable and available often forces people to the urban periphery where units are opening up.

Tax benefits that make public transportation costs deductible encourage and facilitate long distance commuting. The average commuting time for workers in some areas of Japan is 80 minutes each way (Okamoto 1997). Stations in towns and suburbs within 90 minutes of major urban centres experience traffic gridlock in the morning as commuters converge to take trains or subways to work or school, often arriving by car rather than bus. Growing levels of car ownership thus facilitate suburbanization.

The presumption in Japan is that women do not work once they marry and have families. Traditionally women shop every day to buy food for dinner. If they live in the city, they walk or cycle to the shops to buy what they need: small shopping carts in the grocery stores reflect this pattern of daily shopping. Of course, women living in the suburbs need a car to go shopping. If the household has only one car, the wife drives the husband to the train, generating gridlock around stations (Okamoto 1997); if they can afford it, though, households buy a second car.

Suburbs are growing rapidly in Japan. In some cases, these suburbs are former villages that have merged into putative cities. For example, Nisshin (near Nagoya) is a 'city' of 70,000 created from uniting several farming villages increasingly urbanized by the extension of commuter rail and highways past the Nagoya city limits: in effect, Nisshin constitutes a suburb of Nagoya. Smaller towns and cities within commuting distance of the large cities also show many signs of suburban development, in some cases influenced by Western urban theory (see Box 6.1). Much of the development in such suburbs features low-rise detached housing, but some clusters of high-rise apartments also appear, often laced amongst rice paddies, orchards, and factories. Land and housing in these suburbs are cheaper than in the city, and the dwelling units may be larger than those available in the city. With services dispersed over a wide area, households in the suburbs prove more car-dependent than urban residents are. Many households have two cars. In some cases, the suburbs have inadequate services, especially waste water treatment: river systems and wetlands show the negative impacts of inappropriate effluent disposal.

Box 6.1 White Town: American-style suburbanism

In the commuter-shed of Nagoya, Japan's third largest urban concentration, Kisho Kurokawa's firm, Urban Design, designed the *danchi*, or private new town, of White Town in the 1970s. Located outside the city of Tajimi, in Gifu prefecture, White Town is 36 km from Nagoya. Its layout reflects the influence of garden city and Radburn ideas, and a willingness to experiment with a North American grid street pattern in a Japanese suburb. Even in Japanese, the community is known by its English title, reflecting the foreign ideas embodied in the place. It represented a new approach to urbanism in its time, and embodies some of the principles now associated with new urbanism (such as high densities, defined centre and edges, attention to the public realm).

White Town rolls over two hills and across the valley between (see Figure 6.4). Residential pods cap the hill tops, while the central valley has a public corridor with elementary school, convenience shops, apartment buildings, and bus stop. The developer subsidized buses to start transit service in White Town, but service was not especially frequent nor popular when we visited in 1999. The suburb is some 15 minutes by car to Tajimi train station (longer in rush hour) Planners in Tajimi estimate that 80 per cent of White Town's workers commute to Nagoya.

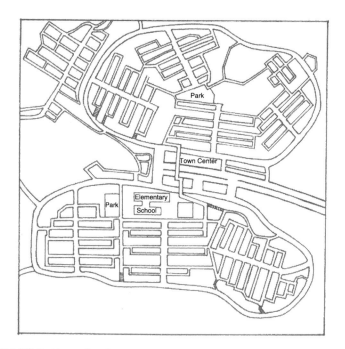

Figure 6.4 White Town site plan
Many of the features pioneered with the garden city and Radburn models appear in the White Town plan. Pedestrian paths make travel safe for children, but force long drives around the ring roads.

Ring roads circle the hills, limiting automobile access to residential neighbourhoods to a few streets. High retaining walls give the impression of a citadel mount. Lots are a minimum of 150 sq m. Internal streets follow a grid pattern, with a slightly different grid orientation for each neighbourhood. Bollards in local streets prevent cars from short-cutting from neighbourhood pod to pod: some of these barriers were original to the plan, and some appear to have been added later. Pedestrians can use sidewalks and paths to cross between neighbourhoods, enjoying a safe travel system to the elementary school: both the developer and residents of White Town praised this safety feature. Although residents say the barriers force a long drive to pick up children in other neighbourhoods, they feel that the pedestrian safety is worth the time incurred.

White Town offers several benefits to its residents. About 6 per cent of the site is attractive park land: high by Japanese standards. Homes are less expensive than houses in Nagoya, offering households the opportunity to acquire more space than they otherwise could afford. The community has the feel of a comfortable family-oriented medium-density suburb: a smaller-scale version of a Western town (see Figure 6.5).

Figure 6.5 White Town street: a Western feel
Although the density is much greater than that found in North American suburbs, in some ways White Town feels transplanted to Japan because of its light-coloured building materials and bay windows. The painted line on the street signifies 'sidewalk'.

Tajimi planners identified several problems with White Town. For instance, early phases did not have enough parking spaces because lots provided for only one car. Illegal parking on the streets has become a significant problem, although a few paid parking lots now accommodate some of the overflow. New homes have space for two cars (the dominant suburban trend today). The planned commercial uses have not been especially successful, and public transportation is not as frequent or rapid as people had hoped. Few jobs are located in White Town. The grid that appears on the site plan functions more like cul-de-sacs on the ground. The planned junior high school was never built, leading to concerns among residents about the distance their children have to travel for education. In fact, children have to go long distances for any kinds of activities. Many residents dread the heavy traffic in the mornings as they try to make their way to the train station. In sum, despite the hope of creating a reasonably self-contained community, White Town has many of the same problems as the typical suburb in the West.

Emulating a Western new town has given White Town a certain cachet, but the model has not inspired imitators. Big development projects like White Town are fairly rare in Japan. Even the developer for White Town has since returned to condominium building in Nagoya. Although *danchi* continue to pop up in Tajimi, most recent ones have many fewer homes and services. Japanese investors who understand what has proven successful in the past may be hesitant about following innovative Western design movements.

Source: Jill Grant and Kaori Itoh

Attitudes to land and housing seem quite different in Japan than those in the West. While the Japanese value land highly and rarely sell it, they treat housing as a consumer good. People view homes as temporary and replaceable despite the high cost of building them. The average lifespan of a house in rural Japan is 26 years: by contrast, a house in Britain lasts 75 years (Kobayashi 2000; Yamaguchi 1999). Zetter (1994) says that 10.9 per cent of Japanese housing units sit unoccupied: this may be because residents have died or moved away for jobs. People rarely rent out or sell off such units. When they do, older buildings can devalue the lot price. For instance, in Nagoya in 1999, I noted that lots sold with houses on them often set lower prices than vacant lots in the same neighbourhoods. The Japanese presume that the new owner will tear down the existing house and build anew, rather than renovate an older home. For this reason, the housing market proves relatively inefficient and arguably wasteful. The concept of sustainability is slow to reach Japanese consumers. Urban strategies that require long-term investment in high quality structures thus face a daunting challenge.

With a public administration long targeted at facilitating economic growth through consumption and development, Japan developed a relatively weak planning system. Power remains highly centralized, with local governments obliged to implement national policy in land use management. The Japanese people were encouraged to accept the development agenda of the post-war period and discouraged from complaining about the externalities that accompany it, such as long commutes, noise, pollution, and limited living space. A strong national consensus supported the modernist project.

Local planning authorities cannot effectively control where growth occurs (Alden and Abe 1994; Hebbert 1986). National planning authorities identified 'City Planning Areas' within which plans designate 'Urbanization Promotion Areas' (UPA) and 'Urbanization Control Areas' (UCA); these designations, though, do not really limit property owners' rights (Sorensen 2002). Haphazard development in small pockets can proceed without services, even in controlled areas. While the idea of designating UPA and UCA seemed a good strategy for managing growth, the execution of the system has been inadequate. Even local governments looking for places for institutional uses like schools and hospitals typically buy sites in the less expensive UCA districts where access to public services like sewerage and public transportation may be poor. Instead of preventing sprawl outside designated planning areas, the system has in some ways facilitated it.

The administrative system in Japan does not encourage innovative or reflective planning (Grant *et al.* 2002). Local administrators in planning offices are rarely trained as planners; they rise to their positions through a civil service system that rotates employees through various departments every few years to ensure that they gain general experience in all elements of the government's business. Hence those writing policy and administering plans often have limited knowledge of planning theory or practice elsewhere. They rarely draw on international examples or experiences, nor do they share their own experiences with planners outside Japan. Planning practice in Japan can seem somewhat insular. Innovative approaches influencing development in other countries may thus have little impact.

With its focus on growth, until recently the Japanese government was not especially concerned with quality of life or environmental quality. The unfortunate result was environmental problems, poor services in suburban areas, little green space, congested cities, and uncoordinated development (Sorensen 2002). In the last few years, though, the government has begun to open opportunities for attention to such matters. Legislation that allows local governments to pass by-laws on planning has spawned a movement for public participation in the planning process. *Machizukuri* ordinances often involve environmental policies and plans; although they lack the force of law, to date they have generally been honoured by other levels of government (Ishikawa 2004; Koizumi 2004). After generations of deference, civil society may be in the process of unfolding in Japan as the *machizukuri* process gives citizens greater influence (Sorensen 2002). If the process continues, that may open new possibilities for more responsive planning for healthier and more adaptive cities. With greater opportunity to voice their concerns, the Japanese may redefine what they consider to be the elements of the good community.

Is new urbanism relevant?

Are new urbanist principles and new urban approaches relevant in Japan? Japanese cities feature many principles that new urbanism advocates. They are high density, compact in form, with narrow streets, mixed use at fine scale, and extensive public transportation systems (Grant *et al.* 2004). It is possible, and even desirable, to live without a car in Tokyo and other major cities like Osaka or Kyoto. We find little evidence, though, that the latest approaches to urban development in Japan respond directly to new urbanism. The major factors driving compact form in the largest cities are still the cost of land, the affluence of the population, and the desire for access to public transportation. Where affluence makes other options possible for households, we see abundant evidence that lower densities, greater car use, and scattered sprawl result. Moreover, rising public concerns about some of the externalities of compact form (such as air and water pollution) could support an agenda of deconcentration (as it did in Europe and North America during the garden city's heyday).

Modernism retains a certain design appeal in Japan. Although some of the wealthy build traditional facsimiles, the lure of classical and vernacular styles has not generated the interest in Japan that it has in the West. Conceptions of beauty and expectations about the character of the urban environment are quite different in Asia than in Europe. Without a tradition of an elaborate and extensive public realm, the Japanese may have little commitment to the kind of policy changes and investments necessary to apply new urbanism principles. A *laissez-faire* attitude (albeit couched in a culture that encourages conformity) proves difficult to shake.

We find little evidence that new urbanism as a theory or practice is having much impact in Japanese thinking about the city and how to plan it, even though Al-Hindi (2001:213) reports that there is a new urbanist project somewhere in Japan. I discovered some areas where people were reviving traditional urban principles in Japan – especially in historic sites that attract large numbers of Japanese tourists. Whether these were indigenous initiatives to sustain an economic resource or design innovations inspired by international movements is difficult to judge (see Box 6.2).

Why has urban form not become a significant issue in Japan? It seems fair to suggest that the Japanese are not architecturally or urbanistically dogmatic. Urban centres and suburbs reveal eclectic and pragmatic tastes. Traditionally, the Japanese have ignored the public realm; hence the parks have been small, public spaces typically unremarkable, and design standards weak.

Shelton (1999) explains that Japanese attitudes towards streets contrast notably with Western ideas. For the Japanese, streets are not important enough to name, regulate, or beautify. They are simply spaces for movement. Instead of using streets for navigation, the Japanese orient themselves by landmarks. Westerners find this strategy an incredible challenge. We pick up a city map expecting to find named

Box 6.2 Hikone rediscovers its history

On the shore of Lake Biwa in central Japan lies Hikone, a university and castle town of fewer than 100,000 people. Tourism has become an important component of the local economy, influencing the redevelopment of parts of the central city by local government and the chamber of commerce. Associated with key events in Japanese history, the town inspires visits from bus loads of school children and elderly people out for day trips. As is happening in such tourist destinations world-wide, redevelopment in ways that celebrate local history becomes an important strategy for marketing the community to visitors.

Authorities have rebuilt the Yume-Kyobashi Castle Road area in Hikone in a traditional style as a castle town. The redeveloped street takes an Edo era (seventeenth-century) theme, while the older downtown reflects the early twentieth century Taisho era (see Figure 6.6). The result has been called 'Old-New Town style' (Yamasa 2004). The reconstruction allows visitors to get a sense of the character and feel of the old Japanese city, while offering new buildings for businesses like restaurants and souvenir shops. Narrow traditional streets make car use difficult in parts of the city: my host had to pull in his car mirrors to drive out one of the lanes from the parking area.

Figure 6.6 Hikone: new traditional urbanism
The main street in Hikone features new versions of traditional buildings designed to capture a sense of history. We might call this a theme-park approach to urban design.

Does Hikone represent an example of the revival of traditional urbanism, or of theme-park historicism? We find similar urban approaches in other historic tourist sites in Japan, like Kamakura. Certainly, Hikone's commitment to its history has a lot to do with its interest in attracting tourists vital to its economy. The project may have a wider significance, though, in illustrating the attractiveness and viability of historic building patterns in the Japanese city. While modernism continues to inspire many architects in the largest cities, a market is developing for traditional urban styles and forms in some parts of the country.

Source: Jill Grant

streets that we will use to make our way around urban districts. Japanese city maps prove utterly hopeless, not only because they are in symbols and a language we do not read, but because in most cases the streets simply have no names. Multifunctional and adaptable, streets achieve purpose and identity for the Japanese by virtue of the activities within them. They need not be 'legible' because their use is limited to those who know them. Small streets connect private residential spaces; strangers have no place there. The concept that the street constitutes an important part of the public realm – a central tenet of new urbanism as a philosophy – is totally alien to the Japanese.

After centuries of living at high densities, the Japanese have developed not only an urban infrastructure that facilitates compact form, but also cultural adaptations to dealing with the externalities of small spaces (Grant 2000a, 2001b). Shutters, bamboo blinds, and frosted glass cover ground floor windows to offer privacy screens. Although the homes are close to the street, no eyes look out, or in. Cultural expectations about appropriate behaviour help to condition civility and conformity amongst the people, so that overt surveillance is not required. Crime rates have traditionally been low, and respect for others high. Walls, fences, hedges, and changes in elevation separate private spaces from public areas: enclosure of private space is the norm. Homes are small, but people use space within them in a flexible way: for living by day, and sleeping by night. High quality in new construction helps to limit noise and odour transfer. Most homes, even apartment units, have outdoor spaces to the south (for drying laundry and airing bedding). The kinds of injunctions that new urbanism projects typically place on hanging laundry or storing materials in private areas, or using front yards to park cars, would shock and dismay urban residents in Japan. New urbanism conveys on urban residents the implicit right to an attractive public realm that includes the parts of private dwellings visible from the public street. Because Japanese culture conceives of no such right, residents can meet their own needs in their exterior spaces.

Western nations can learn a great deal about how to accommodate density effectively from the Japanese experience, but Japanese cities also illustrate the challenges of high density and mixed use. High investments in urban transportation infrastructure are essential to ensuring adequate mobility, but also have the effect of inflating land values and escalating public debt. Long distance commuting results, as households search for housing in their price range. A tight mix of uses can create environmental risks for residents when industrial accidents occur. People pay a price for the hyper-growth that Asian cities have seen in recent decades. Given the alternative (of lower density suburban living with good commuting options), Japanese consumers make choices that facilitate suburban growth.

New approaches to urban development in the West have already offered some lessons for East Asia. A level of cultural borrowing and emulation is occurring, even in nations with divergent economic conditions and political systems. For instance, the interest in public participation and movements for improved environmental quality in Japan and Korea have clearly drawn inspiration from the West. Asian countries have their own hybrid urban approaches that pick and choose strategies to fit their needs and means. While the West looks sceptically at modernism, East Asian urban centres continue their rush to modernize; they have not turned to classical or vernacular design traditions to any great extent. At the moment, given the priorities that drive development in Japan and its neighbouring countries, and regional cultural traditions and values around the urban environment, it seems unlikely that an interest in heritage conservation and traditional urbanism will develop into a full-fledged new urbanism movement as it has in North America and as it aspires to do in Europe. New urbanism may draw on the principles that underlie patterns in Eastern cities, but it cannot speak effectively to the cultural values and behaviours that accompany oriental urban forms.

CHAPTER 7

COLONIAL URBANISM: CANADA SIGNS ON

Eager urbanists

It seems fair to say that Canadian planners have embraced new urbanism. In talks at conferences, Andres Duany suggested that Canadian planning led the way in implementing new urbanism (Wight 1995). In Toronto, Duany criticized Seaside as a failure, but called the Cornell project in Markham 'absolutely flawless' (Warson 1996:21), his 'flagship project' (Bentley Mays 1997:C17). What created the context in which the message of new urbanism would find such a receptive audience in Canada?

Everett-Greene (1997) thinks that Canada proved fertile ground for new urbanism because of its higher densities than the USA (although lower than Europe) and fairly strong central planning authority. In the post-war period, the Canadian government pursued planning as a strategy to promote economic development and industrialization. While never a full welfare state as many European social democracies became in the twentieth century, Canada took an active approach to government intervention in shaping communities. Planning offered the policy and zoning the tools to allow the national government to push an agenda that would have significant implications for local communities.

Planning is easier in Canada than in the USA. Regional governance – and more recently the amalgamation of several large metropolitan areas with their suburbs[1] – gives planners the ability to coordinate policy and regulations quite effectively. Zoning and land use regulations serve as tools for implementing community plans. Accordingly, when planners decide a new policy direction is in order, and persuade decision makers of the validity of the cause, they have the tools to make a difference.

In this chapter, I review the Canadian experience with new urbanism, illustrating its successes and limitations. First I briefly set the stage by describing the historical context of planning Canadian suburbs. Then I discuss some of the larger new urbanist developments and proposals. Finally, I consider some of the challenges that new urbanism faces in Canada.

THE ORIGINS OF THE SUBURBS

In the period immediately after the Second World War, a severe housing shortfall led the Canadian government to invest in facilitating home construction and ownership.

That initiative promoted town planning (Hodge 2003), and disseminated garden city concepts widely (Grant 2003). Government activity at all levels inadvertently supported standardizing the rapidly expanding suburbs. At the federal level, Central (later Canada) Mortgage and Housing Corporation (CMHC) published site planning standards and provided funds for local governments to prepare local plans. It organized home-design competitions and produced books and blueprints of inexpensive homes. Developers bought the patterns and built the same styles of homes from coast to coast (Denhez 1994). Households purchasing homes built according to standard plans on sites that met CMHC standards qualified readily for government-insured mortgages. Municipalities often adopted the CMHC site planning standards as their local regulations, since that ensured that homes built would qualify for government mortgage insurance.

The federal government also stimulated the drive towards mass home-ownership by financing development and building companies to produce housing (Sewell 1993). Eager to create economies of scale to reduce the costs of building, the government provided funds to allow development companies to grow into large corporations, often national in scope. CMHC promoted urban renewal and redevelopment in dilapidated core areas in an effort to modernize Canadian cities. The federal government saw the strength of cities as key to supporting Canada's economic and industrial growth in the post-war period.

Provincial governments also got into the act of suburban development by passing enabling legislation and sometimes providing resources for land banking.[2] Provinces encouraged local governments to hire planners and prepare plans to facilitate urban development.

By adopting land use regulations that showed amazing consistency across the nation, local governments standardized the form of the suburbs. The planning profession converged on principles that ensured a spacious suburban environment with desired attributes of both town and country, based in part on the success of garden city new towns like Don Mills in Toronto (see Figure 7.1). Municipalities across the country settled on suburban lot widths of 15 to 18 m frontage, with 7 to 9 m setbacks from streets. Developers and planners agreed on winding roads and cul-de-sacs as preferred suburban road patterns. Engineers established road standards wide enough to ensure that a fleet of emergency vehicles could respond to any calamity. The garden city fantasy broke down in practice, creating sterile and repetitive suburban landscapes.

By the 1980s, suburbs across the country proved quite uniform, with house types that reflected the decade in which construction occurred: e.g., bungalows in the 1960s, split-entry homes in the 1980s. House sizes increased through the post-war period, with the new suburbs of the 1980s increasingly featuring attached front garages. The large, unused front lawns and picture windows of the 1960s gave way

Figure 7.1 Don Mills: garden city Canadian style
Planned by Macklin Hancock (1994), Don Mills in Toronto was an award-winning community. Emulating some of the features of Radburn, Don Mills established the post-war model of community design in Canada. Its wide lots, deep setbacks, and picture windows were widely emulated.

to wide, paved driveways leading to two-car garages. Domestic life focused on the privacy of the back yard, leaving the front yard as an unused transition space from the street.

CRITICISM OF THE SUBURBS

As residents of a 'middle power', caught between Europe and the United States in social and foreign policy, Canadians respond quickly to innovations and critiques. Cities followed the American trend to urban renewal in the 1950s and 1960s, accepting federal incentives to modernize their cores (Grant 2000b). As battles raged against renewal in America, the same responses emerged in Canada. In the late 1960s, a major spokesperson for protecting older urban districts and pursuing an alternative model of urban development joined the battle in Canada: Jane Jacobs moved to Toronto to protect her sons from eligibility for the Vietnam draft. Immediately joining the 'Stop the Spadina Expressway' movement, she helped to galvanize the forces against renewal, and against urban expressways. Moreover, she became a major influence in the way that Canadian planners thought about urban issues.

Ever the critic of planning, Jacobs insisted that planners take notice of what their policies were doing to the urban fabric. The doyenne of cities, the arbiter of 'great' cities, had chosen Toronto as her home: this reinforced some Canadians' belief in 'Toronto the good'. Jacobs quickly became a Canadian celebrity, working with those who opposed renewal, and insisting that compact, diverse, walkable, and vibrant cities should be the goal. Her book (Jacobs 1961) became one of the sources most commonly cited by Canadian practitioners as an inspiration in their careers (Higgs et al. 1988).

The election of left-leaning, pro-urban factions in municipal governments in Toronto and Vancouver in the late 1960s and through the 1970s (Higgins 1986) created the context in which many planners, politicians, and activists advocated mixed use, high density infill projects to provide affordable housing in the inner cities. For instance, in 1972, a reform council in Toronto revised the downtown plan to encourage an urban character, and to protect neighbourhoods. The city began work on what John Sewell (an alderman at the time) called 'the radically traditional St Lawrence Community' (Sewell 1993:174). The city purchased 20 ha of derelict industrial land only a half-dozen blocks from the financial district and established a citizen committee to plan its redevelopment as an urban neighbourhood. They proposed a model clearly influenced by Jacobs's principles, but modified to suit Toronto. The St Lawrence plan called for medium-high density (40 units per ha) in three-storey townhouses or mid-rise apartments (see Figure 7.2), emulating the traditional Toronto neighbourhood (D. Gordon and Fong 1989). Principles in the design included mixed use, grid layout, traditional building types and materials, human scale, street-oriented buildings, and a variety of tenures (Sewell 1993). Construction began in the mid-1970s. Today the St Lawrence market area is one of the liveliest in the city, with a vibrant mix of uses and people.

Long before Seaside took shape in Florida, the St Lawrence community demonstrated the potential of urban alternatives that would come to challenge the

Figure 7.2 St Lawrence mixed use
This urban neighbourhood in Toronto may have been a prelude to new urbanism. It involved a lively mix of uses within and between buildings, and high density housing for a mix of people.

modernist paradigm. These innovative ideas spread quickly in Canada via professional conferences and journals, and through the active lobbying of individuals like Jane Jacobs and John Sewell. They paralleled interest in compact cities, mixed use, and heritage conservation appearing in the international planning literature.

Other Canadian cities, including Vancouver, Montreal, and Calgary, also promoted infill and intensification policies beginning in the 1970s, but more strongly by the late 1980s (Isin and Tomalty 1993). During the 1970s, Vancouver began redeveloping the False Creek area, a former dump and industrial site along the harbour, with mixed use and affordable housing (Dorcey 2004; Vischer 1984). In the next decade, Vancouver adopted a 'living first' policy that promoted residential intensification within the central city (Beasley 2004). Metropolitan Toronto (1987) published a report urging intensification and infill development as a strategy for accommodating growth in its revised plan process. Planners increasingly rejected the urban renewal strategies of the 1950s and 1960s that had led to commercial development in the central city, and instead argued the importance of bringing people back downtown as part of a 24-hour pattern of uses.

The 1980s added other concerns to the mix of planning ideas in Canada. The energy crisis of the 1970s created fiscal uncertainty and recognition of escalating national deficits. Governments began cutting back on costs, and on programmes. As had happened a few years previously in the UK and the USA, a right-leaning national government came to power in Canada in 1984. While many housing and neighbourhood advocates remained interested in urban revitalization, the focus of municipal planning tended to shift to strategic planning: applying the principles of effective business management to the urban context. Those in industry took advantage of the opportunity to criticize planning as inflexible, accusing it of holding up the market and stifling growth. Many agencies cut back on staffing, putting planners out of work. The crisis certainly encouraged planners to reexamine the basic planning principles that had governed the post-war period, and to find cost-effective development options.

Two other movements in the 1980s contributed to Canadian thinking about planning options and approaches. Following an international conference on health promotion in Ottawa in 1986, several Canadian municipalities began to work towards a Healthy Communities programme. The Canadian Institute of Planners (CIP), the Federation of Canadian Municipalities, and the Canadian Public Health Association convinced the government to fund local initiatives across the country. CIP served as secretariat for the programme from 1988 to 1991, pushing strongly for planners to advocate planning as a tool for promoting community health through effective policy and form (Hendler 1989; Witty 2002).

The Healthy Communities project pushed for a comprehensive approach to health and well-being: walkable cities, meaningful employment, mixed use, access to education and income, clean environments, active living, and disease prevention. It took a public health approach to looking at all elements of life required to avoid illness. Some of its proposals fit well with the principles behind the intensification programmes in Toronto and Vancouver: for instance, the idea of affordable housing options, walkable neighbourhoods, and transportation choices were clearly vital in both frameworks. Most importantly, perhaps, the Healthy Communities project reintroduced the issue of health into the planning vocabulary, and created the framework within which planners could see urban form as contributing to or detracting from public well-being. In some ways, this took planning back to its roots in the public health movement of the late nineteenth century.

Another pivotal paradigm from the late 1980s appeared in the movement for sustainable development. Promulgated by the World Commission on the Environment and Development (1987), the concept of sustainable development had resonance in Canada (Richardson 1989). Rising concern about issues such as ozone depletion, habitat loss, species extinctions, and climate change contributed to a context in which the federal government decided to take action. Sustainable development offered a model that seemed to support economic improvement while also promising environmental conservation and social equity. It became the foundation of Canada's *Green Plan* (Government of Canada 1990), a five-year programme to fund environmental initiatives. Provincial and local Round Tables on the Environment and the Economy soon appeared. As they searched for solutions that might promise more sustainable development strategies, their reports favoured ideas such as the compact city, mixed uses, and enhanced public transportation options (e.g., Nova Scotia Round Table 1992).

Although interest in urban revitalization led to major improvements in urban cores, suburban policies and practices took longer to change. While many cities adopted mixed use zones and reduced lot dimension requirements in central areas, garden city ideas continued to dominate suburban development through the 1980s and into the early 1990s, as developers hesitated about changing development models that met their needs. We can see, however, that strong consensus on implementing principles that new urbanism would adopt was already gaining momentum in Canada well before the American movement convened its first Congress.

NEW URBANISM CATCHES ON

By the early 1990s, many cities were expanding their search for strategies for intensifying development through facilitating urban infill, redeveloping industrial lands, and modifying regulations for suburban areas. Such approaches could be defined as *strategic* (in that they reduced the cost of providing urban infrastructure and enhanced economic vibrancy), *sustainable* (in that they reduced the need for extending development into greenfield sites), and *healthy* (in that they provided for greater diversity and affordability). The Canadian Urban Institute established a regular newsletter, *The Intensification Report*, to encourage communities in their redevelopment processes, and to offer advice on workable strategies for achieving higher urban densities. Government agencies funded studies and demonstration projects to evaluate the merits of new approaches. The various threads of the dominant paradigms of Canadian planning in the 1980s had come together in a revised approach to urban development. Visits to Canada by Duany and Calthorpe in the 1990s found receptive audiences and gave Canadian planners an integrative theoretical paradigm to weave the threads of their developing ideology together. It became clear that Canadian planners had been practising the principles that defined the new urbanism: they joined the nascent movement with enthusiasm (Grant 2003).

With the success of Seaside and Kentlands in the USA, and abundant media attention to new urbanism both sides of the border (e.g., Anderson 1991; Baetz 1997; Barber 1995; Chidley 1997; McInnes 1992; *Newsweek* 1995), Canadian planners, designers, and politicians grew increasingly interested in applying the concept of new urbanism. A professional and popular consensus took shape that new urbanism offered a responsible vision for community design (Chidley 1997; Grant 2003; Wight 1995, 1996).

Through the 1990s, the commitment to new urbanism grew in Canada. Many cities revised their official plans to embed related principles. Zoning regulations were modified to enhance opportunities for mixed use and revised form requirements. Federal and provincial governments promoted the ideas with studies, programmes, and projects (Grant 2002b, 2003; Ontario 1995, 1997). Professional organizations representing planners filled their conferences, reports, and journals with discussions of new urbanism. 'Thus, in the 1990s, Canadian planners practised in a context where New Urbanism dominated discourse and became infused with overtones of environmental responsibility ... New Urbanism (with its rhetoric of sustainable development) provided a reinvigorated rationale for planning practice' (Grant 2002b:75).

Two parts of the country, Calgary (Alberta) and Markham (Ontario), became famous for large-scale new urbanist projects. A rapidly growing city in the foothills of the Rocky Mountains, in 1995 Calgary made new urbanism a mainstay of its planning policy by adopting a transportation plan (Calgary 1995a) and a sustainable suburbs study (Calgary 1995b; White 1996) that clearly reflected new urbanism principles. In the same year, Carma Developments began work on a 'new town' designed in consultation with Duany (see Box 7.1). McKenzie Towne was planned as a neo-traditional development: 12 villages organized around squares, and sharing a common town centre (see Figure 7.3). Narrow streets, garages off alleys, high

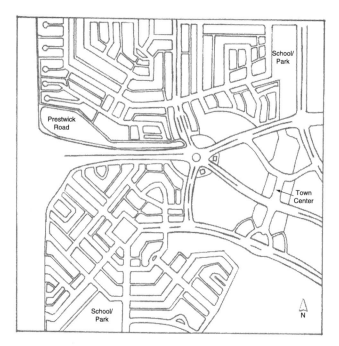

Figure 7.3 McKenzie Towne site plan
Each village in this Calgary suburb was originally planned to have a square or green with narrow roads and alleys. The High Street shopping area, while distant from the early villages, would be central to the project as it develops. As new phases have been built, the original new urbanist concept has given way to a more conventional plan.

density, mixed use, and high design standards characterized the plan (Law 1996; MacDonald and Clark 1995).

Despite the city's hope of enhancing suburban densities through new approaches, McKenzie Towne struggles to meet its targets. Although lots were narrow, small household sizes and greater amounts of roads and lanes mean that McKenzie Towne may not achieve densities higher than conventional developments. Information provided by the City of Calgary (2000) indicates that the average household size in McKenzie Towne was 2.16 persons, while the neighbouring conventional suburb of McKenzie Lake had 3.18 persons per household (see Table 7.1). The good news about McKenzie Towne is that it is more affordable than its conventional counterpart (Calgary 2004a, 2004b). It also has smaller homes and smaller lots. Hence, it consumes less land and materials per household. However, given its small average household size, it actually consumes *more land per person* than does McKenzie Lake. Homes in McKenzie Lake are assessed more highly per area despite the expensive new urbanist flourishes at McKenzie Towne. These data raise questions about the sustainability of the concept of suburban new urbanism in Calgary.

Box 7.1 McKenzie Towne: urbanism in the Prairies

The public buildings and homes in the first phase of McKenzie Towne – known as Inverness Village – proved photogenic and quickly captured the attention of both the mass media and community planners. Before selling a single unit, Carma Developments invested over $25 million in laying out the village square with its gazebo surrounded by elegant brick townhouses and a mixed use commercial building with tower (see Figure 7.4). Given the newness of the concept, the developers wanted buyers to recognize the quality and uniqueness of what the project offered. The early units sold well.

Figure 7.4 McKenzie Towne gazebo
Excellent urban design is the watchword of new urbanist projects. This elegant gazebo in McKenzie Towne is the setting-of-choice for many bridal portraits. Brownstones in the background ring the square.

Located south of the city centre, a half-hour drive from downtown, McKenzie Towne struggles to become more than a far-flung suburb. Narrow roads, sidewalks, back alleys with garages, small set-backs, and narrow lots give the first phase a uniquely urban feel. A school is planned for the neighbourhood. The original 'village' has a mixed-use centre which hosts the post boxes and convenience store; keeping the 930 sq m building fully leased has been a challenge for the developers. Although most observers agreed that Inverness Village is beautiful, recouping the premium required to build such attractive designs proved difficult. In some areas, builders sat on empty units for too long to make the expected return on their investment.

Despite the best efforts of an able developer and supportive city staff, McKenzie Towne faced economic hurdles. For instance, the central 'High Street' has space for community commercial uses. Central to many of the villages planned, the High Street is more than a ten-minute walk from some early parts of the development, and located well off the main access road to the community. Without a volume of drive-by customers, several shops have struggled financially, and some storefronts remained unoccupied for a long period after construction. Although the designers originally hoped to include residential units over the shops in the High Street, the cost of meeting national building codes made such mix prohibitive. On the main street, false second-storey façades replaced promised apartments. While McKenzie Towne achieved a mix of housing types in its early phases, keeping commercial uses viable in a suburban location proved a challenge.

The brownstones and traditional house styles of McKenzie Towne are beautiful, but they do not reflect vernacular Alberta architecture. More commonly, village homes resemble the housing types built in Kentlands, Duany and Plater-Zyberk's project in Maryland. Applying the principles of new urbanism in the Canadian context sometimes involves importing someone else's history. To an extent, the concept of adapting to local context was lost in translation.

Single-detached houses in conventional suburban layouts and housing types have became more common in recent phases of McKenzie Towne. The landscape of the community is pedestrian-friendly, but most residents drive to work in the city centre and shop elsewhere. Bus transportation is available, but with increasing traffic between the community and the central city, the trip takes longer each year.

One objective that McKenzie Towne has succeeded in meeting is that of affordability. The early units in Inverness Village were fairly expensive, but later stages of the project have tended to target the starter market. Although Carma had hoped that a neo-traditional design would give added value to this suburban project, through time it appears to have accepted that the location is most appropriate for a less affluent clientele. Accordingly, the neo-traditional flourishes of the early villages have diminished with each new phase as the builders try to make homes less expensive. McKenzie Towne may have been ahead of its time and place.

Source: Jill Grant

Table 7.1 Comparative statistics on McKenzie Towne and McKenzie Lake

	McKenzie Towne	*McKenzie Lake*
development type	new urbanism	conventional
persons per household (2000)	2.16	3.18
median home assessment (2004*)	$204,500	$237,500
assessment per square foot (0.3 m)	$142	$152
average sized home (2004)	1421 sq ft (127 sq m)	1839 sq ft (165 sq m)
average sized lot (2004)	3865 sq ft (348 sq m)	5179 sq ft (466 sq m)
area of land per person in household	1789 sq ft (191 sq m)	1628 sq ft (146 sq m)

(*Note: Assessment figures do not include condominiums. Amounts in Canadian dollars.)
Sources: Calgary, 2000, 2004a, 2004b

In the 1990s, the provincial government in Ontario and the federal agency, CMHC, proved keen to promote sustainable development. Officials saw new urbanism as a track to sustainability, since the approaches shared many of the same goals and methods. Governments promoted new urbanist design with the publication of volumes of resource materials (e.g., Berridge *et al.* 1996; Ontario 1995, 1997), and with competitions. In 1994, the Ontario government held a design competition for surplus land it had expropriated north east of Toronto for an airport in the 1970s. The Seaton competition resulted in several entries that reflected principles of new urbanism and sustainable development. A few years later, the provincial and local governments hired Duany to help plan Cornell (see Box 7.2).

Duany's influence is clear in Markham, the town that hosts Cornell, and in York Region.[3] Markham had adopted design guidelines for a historic district before the Seaton competition (Markham 1991), but the negotiations over Cornell encouraged town and regional staff to change the way they planned (Gordon and Tamminga 2002). York Region (1994a, 1994b) adopted plan amendments in 1994 to allow two new urbanist projects in Markham: Cornell and Angus Glen. Markham's suburbs from the 1980s reflect the conventional loops and cul-de-sacs of post-war development, but since the mid-1990s new development projects reveal TND style new urbanism prescriptions: straight and interconnected streets, back alleys, limited set-backs, narrow lots, 'traditional' dwelling styles, and attractive public spaces (Gordon and Tamminga 2002). Markham (1998) adopted community-wide design guidelines that promote principles such as grid layouts, structured public spaces, and consistent built form with strong architectural character.

Duany began working with Markham planners in the mid-1990s to create Markham Town Centre on a greenfield site. Created as a political unit from the amalgamation of villages and suburbs, Markham had no original community centre or business district. The plan calls for 10,000 residents and a mix of office and commercial development in the Town Centre over the next few years, all according to new urbanism principles. Other new urbanist projects are being built in various areas of Markham as well. At Angus Glen an upscale traditional community wraps around a golf course. At Berczy Village, a modest development sprouts up houses in rows reminiscent of the corn fields it supplants.

Markham remains strongly committed to new urbanism as a set of coherent planning principles. The planners have taken an active stance to secure a vision of a better community. For example, to ensure that corner stores are not just permitted but are actually built, they will require developers to construct the stores as a condition of permitting development. With a fair degree of success, they have worked to protect ecological assets identified in earlier studies (Gordon and Tamminga 2002). They are learning from the problems they encounter along the way: for instance, that the municipality may lack sufficient resources to maintain the street trees, parks, and

Box 7.2 Cornell: new town on the leading edge

At the northern edge of the Toronto suburbs, where the city oozes into farmers' fields, the new urbanist community of Cornell presents a clear design option to the conventional Canadian suburb. Duany *et al.* (2000:199) call Cornell 'a fairly pure application of the neighborhood concept'. The original plan called for 11 neighbourhoods with reasonably affordable housing (Warson 1996). A total of 10,000 units were planned on a connected grid of streets with homes backing on alleys.

Today Cornell is a vibrant community with active social groups and a strong community identity. The project has sold well in the hot Toronto real estate market. Originally the provincial government had planned to include 50 per cent affordable housing in Cornell, but a change in government saw the land sold off to the private sector for development as the government commitment to low-cost housing disappeared. Although some garages have apartments over them, Cornell does not feature affordable housing. Units are large and well-designed (see Figure 7.5).

Figure 7.5 Cornell: upscale country
Designers of new urbanist homes in Markham used Ontario farmhouses as inspiration for the contemporary architecture. While some homes are modest in scale, others are luxurious. Price tags are similarly impressive.

True to the neo-traditional new urbanist model, Cornell has a mix of housing types in attractive designs. John Bentley Mays (2001:114) calls it a 'quaint pseudo-Victorian package'. My slides of the place look like pastel postcards. Cornell succeeds in its designers' aspirations to make it 'imageable'. Some home styles emulate Ontario farm houses. Others, though, look a great deal like the brownstones of Kentlands or McKenzie Towne, detracting from the originality of the attempt to replicate the local vernacular. Most of the pocket parks prove quite attractive, but the large green in front of the café feels over-sized and unwelcoming.

Cornell's town centre struggled in its early years, but was doing better by mid-2004 (see Figure 7.6). A successful coffee shop mixes with convenience retail and services topped by residential units. Its location away from the main arterial that passes Cornell limits access from higher volumes of patronage. A day care centre recently moved in to take a good portion of the space that had previously been under-utilized. Other commercial areas are planned for later phases.

Figure 7.6 Cornell town centre
The town centre block has commercial uses at ground level and condominium housing above. Given its location within the development, the retail area draws mostly local custom.

In 2003, the Town of Markham adopted strict architectural standards for Cornell to ensure quality. 'In creating new buildings in Cornell, care must be taken to ensure that their designs follow consistent principles and do not rely on false replications of "olde" styles through the application of insufficient detail or inaccurate reproduction' (Markham 2003:3). The guidelines indicate the influence of Duany's notion of the transect: 'Neighbourhoods along the ninth line and internal to the community are intended to be more urban in character, while neighbourhoods located along the north and east perimeter adjacent to open space reserves are to be less urban in form' (Markham 2003:3).

Duany remains on retainer to the Town for advice on Cornell. A planner for Markham told me in 2004 that Duany usually comes in each year to make sure the project stays on track (at $500 an hour, paid by the developer). The initial developer, Law, ran into financial difficulty and has sold out. Planners note that they have to battle each new developer and builder to maintain the original vision. For instance, developers have resisted alleys because of the cost implications. The alleys have presented problems for garbage collection and snow plowing; because they are private roads, the municipality does not service them. Residents resented the inconvenience of taking their garbage to the end of the alley, and digging out their cars after a snowfall. Planners work diligently to hold developers and residents to the initial promise of Cornell.

> Although a new town in Markham was originally conceived as a model of sustainability and responsibility, as built Cornell represents another upscale suburb. Homes in Cornell are quite expensive (as is true throughout the region), with narrow townhouses selling for $220,000 in the summer of 2004. The design guidelines call for quality measures that drive up the cost of housing. While buses do service Cornell, most households remain car-dependent. A Markham planner who led a tour of Cornell at the 2004 CIP conference described the community as 'transit ready'. As of yet, the train line does not extend all the way to Cornell. The development has few jobs and limited commercial uses. It is certainly an attractive suburb: perhaps the best of the Canadian suburban new urbanist 'proto-towns'. Given its attention to high quality urban design and environmental conservation, and its location in the Toronto periphery, Cornell seems likely to mature well and retain its value through time.

Source: Jill Grant

squares required in new projects, and that snow storage and emergency access present challenges in areas with narrow roads and lanes in winter.

Several other Ontario municipalities have also encouraged developers to experiment with new urbanist projects. The City of Waterloo (1998) adopted a 'west side vision', and approved several new urbanist-inspired projects such as Eastbridge and Clair Creek Village. New urbanist projects also appeared in places like Oakville, Orangeville, Niagara-on-the-Lake, and Windsor. In 1997, the Ontario government released a *Provincial Policy Statement* that advocated cost-effective development patterns, and called for concentrating growth in existing cities, towns, and hamlets. The policy reflected many principles congruent with new urbanism and the emerging paradigm of smart growth. Reports sponsored by the federal government (e.g., Peck and Associates 2000) argued for similar policies, calling the resulting communities 'sustainable'.

By the late 1990s, several urban infill projects adopted new urbanism principles for redeveloping former industrial or military lands. Garrison Woods, on a redundant military base in central Calgary, proved more successful than the earlier greenfield site in the Calgary suburbs. The community, only ten minutes' commute from central Calgary, offers a mix of rehabilitated military housing units and new homes in a compact grid layout (see Figure 7.7). By virtue of its location, this neighbourhood links directly into the urban transit system and offers viable alternatives to automobile use. Its commercial component connects to a nearby commercial core. Garrison Woods has built effectively on infrastructure of the city around it (Bergen 1998), and has proven successful.

To dispose of or redevelop surplus federal land, the government created an agency called Canada Lands Company. Its success with the old army base in

Figure 7.7 Garrison Woods: urban rehab
Former military housing has been renovated to look 'traditional' in this Calgary infill project.

Calgary has led it to take new urbanist principles around the country to other areas with lands being recommissioned for urban use. Many former bases, often originally designed as garden suburbs, are now taking new forms as urban villages.

The same ideas are influencing redevelopment in private projects on redundant industrial lands and sporting venues. In Toronto, the site of a Massey Ferguson tractor works has become a townhouse development, and the old Greenwood Racetrack site has been redeveloped as 'The Beach', a community of old-fashioned houses (Bentley Mays 2001).

Increased flexibility in regulations for The Kings district in Toronto assists former industrial areas to accommodate a new mix of uses that includes housing and retail (CMHC 2004). Heritage factory buildings are finding new uses as offices and housing. High design standards in the district will facilitate the addition of at least 7000 residential units, mostly households without children.

Toronto has approved a proposal to renovate some of its public housing at Regent Park to create a new mixed income community. Regent Park represented the height of modernist ideas for urban renewal when constructed in the 1950s and 1960s. Like HOPE VI in the USA, the idea of the proposed programme is to replace aging public housing units with a mix of subsidized and market units in a new urbanist street configuration. The proposal would double the number of dwelling units on the site, with half sold to new residents. Sales of the market units would help pay for the subsidized units, according to the Toronto Community Housing Corporation CEO. Douglas et al. (2004) worry that many displaced tenants will not get their homes back if the project proceeds; they fear that gentrification in well-located neighbourhoods will force the poor out to the suburbs.

A brownfield site in Montreal has become the locale for developers trying to create a 'signature town': the Bois Franc neighbourhood (Sauer 1994). Its designers found inspiration in the plan for Savannah, Georgia, with small residential blocks structured around squares. Principles derived from urbane French-speaking communities may prove sympathetic to new urbanism: several Montreal designers, like Avi Friedman (2002), have become strong advocates of related ideas.

By contrast with the neo-traditional models of new urbanism practised in many Canadian cities, Vancouver has engaged in what many have called a modernist version of new urbanism. Vancouver's urban design staff have established high standards for density and design in the central city. Langdon (2003c:8) suggests that 'Vancouver could be North America's biggest demonstration of New Urbanism's ability to adopt a distinctly modern architectural expression'. Vancouver is mixed-use, socially diverse, and compact: 50 per cent denser than Metro Portland, and 100 per cent denser than metro Seattle. Olson (2002:55) calls Vancouver a 'unique manifestation of new urbanism: denser, taller, and without the historical pastiche'.

Vancouver planning co-director, Larry Beasley (2004), told the CNU congress in Chicago: 'Vancouver's recent development, especially in the inner-city and surrounding neighbourhoods, offers a text-book case of the Charter of the New Urbanism'. While Vancouver prefers modern architectural expressions, its policies insist on compact, intense, and mixed uses. The city accommodates high-rise towers but requires attention to the public realm with trees and boulevards. Retail uses are required at ground level on commercial streets, with townhouses on residential streets. Laws protect sunlight and views. Vancouver has some of the most active community programmes for planting and community gardens in Canada (Duncan 2003; James 2003). A strong public participation process ensures community buy-in.

Vancouver enjoys the reputation of a livable, albeit expensive, city. As Beasley (2004) notes, planners have significantly increased the number of people living downtown, reduced the number of cars commuting into the city, and started a baby boom downtown. The city rezoned Pacific Place, the site of the 1986 World Exposition on False Creek, from industrial use to housing after it became vacant. When complete, the site will house 47 slender towers and townhouses for 18,000 people: a kind of urban village. Vancouver requires developers to design 25 per cent of units for families with children. The mix of income levels must include 20 per cent non-market housing. Projects like these go a long way to accommodating urban growth within the existing urban framework and demonstrating the potential of new urbanism principles to offer a viable framework for development.

The latest proposal in Vancouver is for Southeast False Creek. The plan for a sustainable urban development on former industrial lands along the harbour will serve as the athlete's village for the 2010 Olympics (Mallet 2004). The project targets reduced energy use, green buildings, high amounts of parkland and habitat protection, 35 per cent family housing, and 20 per cent non-market housing. New urbanism principles adopted for the project include a mix that favours daily shopping, employment opportunities, transit orientation, and live/work units (Vancouver n.d.).

In sum, new urbanism has had a considerable impact on planning in Canada in the last decade. It has encouraged planners and decision makers to reduce development requirements in ways that permit smaller lots and setbacks not only in new urbanism projects but even in conventional suburbs. Minimum lot sizes have decreased in many areas. While 15 m frontages proved standard in the 1970s, by 2004 we more commonly see minimums of 10 to 12 m. Canadian cities have returned to pre-automobile standard urban lot widths. Urban densities are increasing in Toronto (Bourne 2001; GHK Canada 2002), the third-fastest growing city in North America (Lorinc 2001). The largest cities, like Vancouver and Toronto, and those growing rapidly, like Calgary, favour high density development in infill projects. New urbanism provides a theoretical justification for increasing urban densities, for encouraging the mixed use zones planners have applied since the late 1980s, and for promoting urban revitalization and infill growth.

The market for new urbanism seems quite good, especially in areas with strong growth, like Toronto and Vancouver. New urbanism adds value to small lots by giving character and charm to new suburbs and infill development projects. Traditional housing forms present a strong lure for buyers. Bentley Mays (2001:120) quotes a resident of a Toronto new urbanism project: 'Maybe we're on a *Pleasantville* set, maybe we're actors in a movie. But we like it. Our life is our movie.' Maybe homeowners are consuming an image of community, or maybe, as Bentley Mays puts it, people are just trying to 'get back home' (2001:120).

Smart growth has also become popular in Canada, augmenting the visioning exercises so popular in the late 1990s (Shipley 2000; Shipley and Newkirk 1998). Ontario established regional Smart Growth Panels to make recommendations to the provincial government. British Columbia has a network promoting Smart Growth. Cities have also signed on. For instance, Smart Guelph (Guelph 2004) prepared a report that articulates new urbanism principles such as intensification, grid streets, alternative development standards, transit supportive densities, mixed-use development, and pedestrian orientation. Halifax initiated a regional planning process under the rubric of 'healthy growth' (HRM 2001): a rose by any other name. The report of a government party task force (Liberal Party of Canada 2002) has helped to place urban issues on the national political agenda in a way that favours smart growth.

Canadian new urbanism followed generally in the American model of New Urbanism, with some adhering to TND approaches and others favouring the TOD philosophy: some also use the language of urban villages, but less commonly. Duany and Calthorpe have played major roles in offering advice and guidance to Canadian planners and decision-makers and have thus helped to shape the perspective. We can also see, though, that the Canadian practice has reciprocally influenced the American New Urbanism model. Whereas in the USA new urbanists relied on developers to apply codes and covenants, in Canada municipal governments showed that they could achieve the same principles by adjusting planning policies. Whereas in early American new urbanism projects neo-traditional architecture abounded, Canadian cities showed that new urbanism principles could work with contemporary architecture and high-rise structures. The Canadians involved in the discourse about new urbanism helped to raise the profile of concerns about sustainability. In sum, the success of new urbanism ideas in Canada helped to shape the North American New Urbanism approach and agenda.

Challenges to new urbanism in Canada

While policy and plans in many parts of the country have adopted new urbanism principles, practice may prove more resistant to change (Grant 2002b, 2003). Although a growing number of consumers find traditional home styles attractive, many home buyers remain quite happy with less expensive conventional garage-front houses. Some new urbanist embellishments have been integrated into the suburban aesthetic. Hence we often see front porches, heritage-inspired windows, pitched roofs, and gingerbread trim in the suburbs, even on houses with protruding front garages (Grant 2002a). Except in rare cases, this *faux* new urbanism is bereft of social objectives, mixed use, or transit availability.

New urbanism has helped to make the lack of affordability in the suburbs an issue. The Canadian experience to date indicates that new urbanism principles have not remedied the problem. In general, the suburbs with the strongest and most comprehensive new urbanist features also feature amongst the most expensive. Design guidelines (e.g., Ajax 1999; Markham 2003) drive up the cost of traditional developments by calling for high design standards and design architects to monitor plans. Authenticity does not come cheaply. Developers say that new urbanism costs more to build than does conventional suburbia. It also appeals to a different market segment. In Canada today, new urbanism tends to target the upscale market.

The new urbanist projects built to date have made relatively little difference in car use. The greenfield projects are remote from employment and commercial centres: transit ready, but not transit-oriented. Infill projects have proven more successful in creating residential environments where walking and transit use are convenient and viable. Thus we see that urban redevelopment makes sense in achieving the transportation objectives of new urbanism, but in most cases greenfield development offers little hope of making a difference.

Although new urbanist projects have met some environmental objectives in areas like Markham (Gordon and Tamminga 2002), greenfield projects clearly result in a loss of landscape function and often of agricultural land. Planners are hard-pressed to call New England brownstones popping up in farmers' fields a 'sustainable' urban form. New urbanism projects in the suburbs of Toronto, Waterloo, and Calgary are replacing good quality farmland: an asset in short supply in Canada. Achieving sustainable development through new urbanism will have to insist on a shift from greenfield to brownfield redevelopment for long-term viability.

We might also question whether the greater building density in new urbanist projects indicates sustainability, or whether it reflects greater consumption of materials per person than did the construction of the 1960s. While we see smaller lots in many parts of the country, we also find bigger houses. CMHC reports that the average house today has one more room than houses had in 1961 (up from 5.3 to 6.3 rooms); houses today are twice the size of the average 1961 house. At the same time, household size has shrunk from 3.9 to 2.6 persons (Statistics Canada 2004). Fewer people consume more housing, even at higher dwelling unit densities.

We might argue that if the idea is to conserve space, then maybe a skinny house with a ground floor garage (see Figure 7.8) is more 'sustainable' than a house

Figure 7.8 Skinny house: an excuse for ugliness?
This detached house in York Region, north of Toronto, is on an extremely narrow lot. To conserve space, the first floor is primarily garage, with living areas above. The result is not beautiful, but it is relatively affordable and it optimizes suburban densities for detached housing.

with a garage on a lane. The combined building avoids the larger footprint taken by a house, outbuilding, and lane while conferring the advantages of a single detached house. It thus leaves more usable lot for outdoor living, gardening, or wildlife habitat. Where York Region allows some very narrow lots, developers can achieve net residential densities of up to 25–30 units per hectare with detached homes. The resulting product may not be beautiful, but is ugliness sufficient justification to avoid particular housing forms? Which do we value more in a good community: affordability, environmental responsibility, or beauty?

While a mix of housing types occurs in communities where land costs are highest, mix seems rare elsewhere. Consumers continue to resist mixing of household types and of non-residential uses. Sensitive to the issues that mixing can create, developers tend to promote mix primarily in urban redevelopment districts. Accordingly, relatively little mixing occurs in suburban areas. Moreover, even when housing types are mixed, the residents in particular neighbourhoods tend to come from similar backgrounds and aspirations. Social mix has proven hard to achieve.

Flying into the airport of any Canadian city reveals that developers continue to build single-use suburban areas on cul-de-sacs. Homes are often set back from the street with attached garages for the affluent, or front driveways for those of lesser means. Alleys appear where required by local regulations designed to encourage new urbanist approaches, but seldom elsewhere. Climate may be an issue for those who hope to sell houses on alleys, because of the difficulty of managing snow removal. Harsh Canadian winters make attached garages especially attractive to consumers anxious to step directly into warm vehicles on cold February mornings.

Some parts of Canada – the largest cities and energy-rich areas – are growing quickly. Most of the country does not enjoy such rapid expansion. New urbanism touches in site planning prove hard to sell in slow growth or no growth areas. In those communities, large lots remain possible and popular: a sign of success and economic vigour. Local officials have no incentive to modify regulations to promote new urban approaches. Declining communities cannot afford urban designers on staff to encourage developers to experiment with innovative concepts. Even if planners in such communities agree with the principles of new urbanism, they cannot implement the ideas.

New urbanism has encouraged cities to focus on the public realm. Hence suburban Markham, aspiring to become urban, is developing a Town Centre. Many municipalities have engaged with urban design in the last decade and are using new urbanism principles to revive their downtown cores and waterfronts, or to create urban villages. For others, though, the resources required to succeed in new urbanism have proved difficult to find. Municipalities need money to build parks, to repair declining infrastructure, or to improve public transportation systems. Canadian municipalities typically lack sufficient taxing authority to raise funds to make the kind of urban places to which they aspire.

As in other post-industrial societies, increasingly the private sector in Canada provides alternatives to the public realm. Town centre malls and entertainment districts attract affluent consumers in their recreation hours. Upscale suburbs have beautiful private squares, greens, parks, and pools, while affordable suburbs sometimes have little quality public space. Municipalities that lack funds to equip parks may prefer to take cash donations from developers instead of insisting on new playgrounds and ball fields that they cannot maintain.

The largest cities, Vancouver and Toronto, have taken some initiative to force or entice developers to provide affordable housing. By and large, though, municipalities do not push developers to build non-market units. Achieving goals of equity and affordability remains a dream in Canada, but not a reality of planning practice in new urbanism projects or in other developments.

New urbanism is certainly making many neighbourhoods more attractive. Can it also make them better communities? Some of its proponents think so. For instance, the design guidelines for Markham (2003:4) say: 'The creation and integration of many urban parks . . . will complement the people-friendly streetscapes by providing spaces for human interaction, and thus, enhancing a sense of belonging to a community.' It may seem a slight stretch to suggest that parks will enhance community, but they certainly can improve tax revenues and attract home buyers. As long as the market responds to new urbanist touches, we can expect developers to continue to use them.

THE FUTURE OF NEW URBANISM

How can we explain why an American idea had such resonance in Canada? The new urbanism articulated values and principles that had already gained credibility in Canada during the 1970s. It gave voice to the beliefs that many influential planners and designers had been practising without knowing what to call their ideas. In other words, new urbanism represented a theory that could explain and legitimate planning practice as it had been developing.

How has new urbanism been transformed in practice in Canada? Although some Canadian planners may identify what they are doing as new urbanism at meetings of the Congress for the New Urbanism, I find few explicit references to 'new urbanism' in the many plans and policies I have examined. The principles of new urbanism are now discussed in Canadian practice as simply 'good planning principles'. They have effectively replaced the earlier garden city planning principles in professional reports, conferences, and discussions. Many cities have adopted design guidelines and modified land use regulations to encourage or facilitate the application of new urbanism principles. The values and premises of new urbanism are now described as ideas that contribute to healthy, sustainable, smart, and livable cities. It is difficult to find a planner who does not believe in mixed use, compact development, or transit-oriented design.

Of course, pre-existing factors and consumer behaviour remain important in practice. With a network of highways and high levels of car ownership, Canada shows no reduction in car use. Although transit ridership is high in the larger cities, most Canadians drive to work from their homes in the suburbs. While some developers are finding success with new urbanist development formats, most remain wedded to conventional development forms that have constituted their bread-and-butter for decades. Consumers who barely qualify for mortgages are happy enough to buy modest production homes in ordinary suburbs. The big box retailers show no signs of changing their practices.

Despite their failings, post-war government policies, development economics, and planning principles kept home ownership achievable for 65 per cent of Canadians. That the well-intended garden city ideals degenerated over time into what many describe as a 'sterile' anti-urban form undoubtedly reflects the operation of economic processes whereby producers head for the lowest common denominator the market will bear. Only at the point in which criticisms of the result become overwhelming does the system respond with a new model. That has happened with new urbanism, at least for some of the players. Whether new urbanism supplants the influence of the garden city model over time remains for the next generation to evaluate.

Contemporary interest in the city is strong in Canada. The national government has promised a new deal for cities which mayors hope will include funding to update infrastructure and to encourage new development models. This agenda provides an opportunity for planning and planners to make a difference at a time when new urbanism is seen as having something credible to offer. Canada may remain the bright spot for new urbanism for some time yet.

PART 3

THE PROSPECTS FOR NEW URBANISM(S)

CHAPTER 8

RECONCILING NEW URBANISM'S THEORY AND PRACTICE

EVALUATING THEORY AND PRACTICE

Now that we have described the practice of new urban development in several regions, we can begin to evaluate practice against theory. Do contemporary practices live up to the new urban agenda? The lessons of practice reveal the challenges faced by those who hope that new design strategies can resolve old planning problems. Moreover, exploring practice exposes the inability of new urbanism (like the planning paradigms that preceded it) to satisfy many of the key concerns that planning theory suggests the good community should address. While new urbanist developments in many countries emphasize the potential of the movement to create beautiful and urbane built environments, we find little evidence that these communities meet the complete range of attributes of good communities.

Does new urbanism represent a 'social synthesis' as Krier (1991:119) would have it, or is it merely the latest example of physical determinism presented by a profession especially prone to simplified notions of how to create positive social behaviour (Harvey 1997)? Certainly, the advocates of new urban forms slip into 'messianic assertions about the important role of good physical design in making a good society' (Pyatok 2000:810). The work of designers involves manipulating space: that may make them prone to generating theory to affirm their beliefs that their actions are effective, both in an aesthetic and a social sense. Planners may prove slightly more cautious, but they increasingly see new urbanism as a means to desired ends (Garde 2004). Practitioners often succumb to deterministic language and expectations.

By contrast with the practitioners, the academic proponents of new urbanism, such as planners Chuck Bohl (2000) and Emily Talen (2000b, 2000c), present a more cautious framework for explaining the relationship of design and social behaviour. They suggest that good design can create contexts that support positive social environments. Instead of environmental determinism, then, they speak of environmental affordance. Talen (1999, 2000a, 2001) writes extensively to persuade new urbanists to refrain from suggesting that their designs condition behaviour, and to convince sceptics that new urbanists do not claim that subdivision layouts change behaviour. By and large, however, her efforts do not seem to have persuaded the leaders of the movement to tone down their rhetoric. The architects in the movement seem perfectly comfortable with linking behaviour to design, while the planners and geographers treat such claims with scepticism or scorn.

In their influential text, *Suburban Nation*, Andres Duany, Elizabeth Plater-Zyberk, and Jeff Speck (2000) make their position clear. 'To begin with the obvious, community cannot form in the absence of communal space, without places for people to get together to talk' (2000:60). While one might argue that statement reveals a position of environmental affordance, their next claim seems more deterministic: 'it is near-impossible to imagine *community* independent of the town square or the local pub' (2000:60). New urbanists suggest that people need a public realm of streets, squares, and parks to meet others who are different from themselves. In this view, community-shaping social interaction occurs primarily in the public realm, rather than in places of work or domestic engagement, and in a context of difference rather than similarity. The designer plays a pivotal role in providing the receptacle spaces – the physical containers – that allow such integration of difference to develop into community bonds. For new urbanism, in some sense, social networks are products of space.

Our review of new urban approaches in several regions has shown that some principles are commonly held as appropriate to good urban form in the early twenty-first century. For instance, we find widespread consensus around the significance of mixing uses and integrating housing types and tenures; encouraging compact form and higher densities; and facilitating public transportation and walking options. Specific design questions, like architectural styles and street patterns, generate diverse responses in particular contexts. While traditional architecture is popular in Florida and central Canada, modernist forms drive new urban approaches in Germany and Vancouver. The grid is great in North America, but the cul-de-sac and mews have kerb-side appeal in the UK. Even the principles held in common face challenges in practice, as they prove difficult to implement to the extent that designers might envision.

Certainly the North American 'New Urbanism' is by no means universal, despite the claims of some of its adherents about its spread world-wide. Even in the United States the movement is not completely unified, but includes competing ideas about how to resolve questions of accessibility (both for cars and for those with mobility impairments). Definitions of the good community seem intrinsically linked to local context, history, and culture, and to the individual interests of community residents. Finding universal principles has not proven easy in a context where issues keep arising, problems prove intractable, and differences will not disappear.

The prescriptions of new urbanism have generated challenges about their universality. Not every city can emulate the animated streets of Barcelona. We see incredible diversity in opinions about architectural style, transportation solutions, and street patterns in the approaches to new urbanism used in various parts of the world.

For instance, while modified grid layouts and front porches have dominated North American new urbanism solutions, they have not proven as popular elsewhere. Rather, designers have turned to local antecedents to find design approaches that resonate with people's understanding of tradition and their expectations of life in the contemporary city. That diversity and complexity makes the cities of the world interesting. While new urbanism acknowledges diversity, its overwhelming preference for identifying universal principles and solutions might suggest that it seeks to reduce or deny difference or 'otherness' (Till 1993). Al-Hindi and Staddon (1997) say that neo-traditional planning talks about diversity but does not practise it. Despite calls for flexibility and adaptation, we can see (especially in the North American new urbanism) a potential for homogenizing strategies: what Al-Hindi and Staddon (1997:369) call the 'totalising undercurrent' of the movement. Certainly the similarity in the look and the patterns in many North American new urbanist projects dispels the promise of variety.

The traditional or neo-traditional elements of new urbanism reveal a deep need for security, image, and identity in the contemporary world. Rapid social and technological change has helped to generate longing for landscapes with symbolic meaning, especially in societies troubled by declining civility and endemic inequality. Empirical studies indicate, though, that what urban village advocates may define as lively pedestrian precincts, lower-class inhabitants may see as a landscape of poverty and disadvantage (Biddulph 2003b). Being able to walk to the store to buy milk is only a positive feature of a neighbourhood when people do it out of choice rather than necessity. In other words, the meaning of landscapes is socially constructed through the symbolic interactions of those who occupy them (Berger and Luckman 1966; Blumer 1969; Graham and Healey 1999). Designers inscribe their own values on urban forms, but the people who use space may read different messages that suit their needs and circumstances.

In this chapter I consider some of the key theoretical claims of the new urban approaches and explore whether practice to date indicates that these claims have been or can be realized.

THE TRAJECTORY OF NEW URBANISM(S)

The new urbanists, like the advocates of the garden city and the modernists before them, began with great hopes about meeting the needs of the dis-empowered and restoring the beauty and character of the city. Their critics might argue that instead of saving the city new urbanists have ended up serving development interests. The garden city in its popularized form came under attack by the 1970s for generating sprawl and monotonous suburbs. Modernism carried the blame for generating ugly places with no character or soul. After 25 years of practice, new urbanism finds itself accused of promoting gentrification, enabling attractive sprawl, and reducing the availability of public housing (Pyatok 2000). Instead of creating real communities, some allege that new urbanists have built illusory Potemkin villages (Shibley 2002), or stage sets for 'reality' entertainment. New urbanism may generate beautiful places with no heart.

The widespread interest in new urban forms in many countries undoubtedly reveals consensus about the problems of sprawl and the need for new solutions. The post-modern city searches for a sense of place, community, and identity while accepting a level of uncertainty. We also find in recent theory and practice a convergence of rhetoric around options described as 'sustainable', 'smart', 'livable', or 'healthy'. The application of these concepts in projects has changed the way places look, especially in North America (Al-Hindi and Till 2001). The revival of classical principles of design has created a new urban aesthetic in areas with relatively recent urban traditions.

Can new urbanism resolve the problems of the city? Its definition of the issue is simple and all-encompassing: sprawl is the villain. The solution seems equally facile: new design codes to eliminate sprawl and obviate its anti-social characteristics. Such simplicity, however, ignores the complexity of the problems of contemporary cities. It also sidesteps the variety of particular local issues that may require addressing, instead opting for universal solutions.

Many common trends remain worrying, and limit the potential for new urban solutions to make a difference. Overall car use shows no signs of decreasing, even in Europe and Asia where gasoline prices are double those of North America. Indeed, China has set targets to expand its levels of car ownership dramatically in the next decade. While public transportation use is increasing in some cities, it occupies a small share of the modal split in most North American cities, and may be declining in some areas where it has traditionally been the principal option. Reducing the environmental impacts and social consequences of car use will take more than community design strategies.

Urban densities are up in high growth areas, where costs drive efforts to conserve space, but not in regions where land costs are modest. Efforts to change historic density patterns run into cultural roadblocks. Overcoming the stigma of high density housing remains a significant challenge, especially in North America.

Asian cities have a legacy of high densities, but as income increases urban residents often look for a greater amount of domestic space. Rising incomes fuel suburbanization, land fragmentation, and long distance commuting (Marcotullio 2004; Sorensen 2004). Ironically, suburban landscapes are often developed in clusters at high densities, but at considerable distance from contiguous urban development: thus we witness high density sprawl.

European cities generally enjoy high densities, and a great deal of redevelopment in the cores is occurring at high density (De Roo 2004). High land costs continue to make compact form the option for housing for most residents in Europe. We also note, though, that households that can afford lower density options often select those choices, especially if they have young children.

While Asian and European families have adapted to urban apartment living, that is less true for households in North America. Higher density urban living in North America tends to accommodate professional singles and couples, the elderly, or the poor. Cultural traditions of spacious living and raising children in suburban environments will prove difficult to change, and may require other policy and financial incentives. Good design is a necessary but not sufficient condition to generate high density living in North America.

The contemporary urban environment presents significant challenges to those seeking urban solutions that create good places. The gap between rich and poor shows signs of widening. Globalization does nothing to reduce – and may even exacerbate – homelessness, underemployment, urban migration, and rural decline. Instead of creating the good communities anticipated by new urbanism, developers are cherry-picking the design elements of neo-traditional urban design (see Figure 8.1), especially the front porches and historical flourishes on small lot housing (Grant 2002a). The radical elements – like mixing uses, classes, and races – are generally excluded in the wider market place where they may be perceived to generate too much risk (Gyourko and Rybczynski 2000, 2004).

Figure 8.1 New urban redux
A suburb in Halifax Regional Municipality NS has homes with traditional flourishes (pitched roofs, porches, and bay windows), but in a conventional suburban situation. Like the garden city before it, new urbanism may be reduced to a few essential elements by developers who isolate the saleable features and market them relentlessly.

DOES NEW URBAN PRACTICE LIVE UP TO ITS THEORY?

In many ways, theories about urban form and planning associated with new urbanism became mainstream ideas in the late twentieth century. New urban principles became the espoused theory of the planning and community design profession. Hundreds of developments reflect the principles and contribute to the spread of interest in related approaches. How has practice responded to the challenges? Are the projects built achieving the key values of new urbanism? Are they affordable, authentic, democratic, diverse, equitable, and sustainable?

ARE THE DEVELOPMENTS AFFORDABLE?

Proponents often justify new urbanism by claims that it can provide more affordable housing by using less land and mixing unit types. In practice to date, however, few new urbanism projects have proven affordable. In fact, several studies reveal evidence of a premium for new urbanist development in the market-place (Eppli and Tu 1999; Gause 2002; Song and Knaap 2003; Tu and Eppli 1999). Even the organ of the Congress for the New Urbanism boasted to its readers about the $24,000 premium on units available at Orenco Station in Portland (*New Urban News* 2003b).

Of the communities we have described, few offer affordable housing. While new urbanism reveals a genuine concern with providing a quality living environment for all, practice has shown that the private sector has trouble providing affordable housing: as Bohl (2000) says, affordable housing has to depend on the public sector. The projects with high proportions of public housing in Europe rely on government or charitable financing to reduce the costs to prospective tenants. They usually include a high proportion of apartment units and row or terraced housing. Vancouver has increased affordable housing supply by a public policy of requiring developers to include 20 per cent non-market housing in key areas showing that with political will, governments can transfer responsibility to the private sector; however, such requirements affect housing prices and the marketability of market units (Brophy and Smith 1997; Neuman 2003).

The new urbanists' participation in the American HOPE VI public housing renovation projects has generated considerable debate about the role of new urbanism in supplying affordable housing; HOPE VI destroyed many more low-cost units than it generated, rendering the new urbanists vulnerable to charges of diminishing affordability and housing supply for the poorest households (Goetz 2003; Marcuse 2000; Popkin *et al.* 2004). Certainly the American new urbanists express little interest in changing the economic or political *status quo* to improve income of the poor so that households can afford to find housing in the market; perhaps, as Ellis (2002) suggests, it may not be fair for us to expect them to. Instead, designers have generally seemed content to work within the market paradigm that generates large numbers of high-end housing projects but few low-cost options. Of the private new urbanist projects discussed in earlier chapters, only McKenzie Towne in Canada proves less expensive than its conventional counterpart; its cost savings came with adjustments to design quality in later phases of the development. For the most part, new urbanist projects have done well in the market place and have consequently become less affordable with each passing year. New urbanism thus has trouble delivering affordability (Garde 2004). The good community will require equitable income distributions, adequate financing, or government intervention as well as excellent design.

ARE THE DEVELOPMENTS AUTHENTIC?

The new urbanists often write and talk about 'real', 'true', or 'authentic' neighbourhoods and communities. They see contemporary cities as false examples, missing the essential characteristics of urbanity. Duany, Krier, and other neo-traditionalists are especially prone to demand authenticity in new urban forms. In this they differ from the proponents of transit-oriented development and urban villages, who more often use the rhetoric of sustainability or smart growth in their writings.

Certainly those who live in the new urban communities view them as real. People raise families, make friends, and enjoy their lives in places like Kentlands, Celebration, and Poundbury. They experience commonality and belonging in these places (e.g., see Frantz and Collins 1999). Communities that photograph well cannot be chastised purely on aesthetic grounds if they provide socially-meaningful places for their residents, any more than suburbs that look abysmal in photographs should be said to deny their occupants meaningful home spaces. Real communities come in many forms.

The critics of new urban approaches, however, often deny that the created settings of new urbanist developments constitute real places. They say the aesthetic touches mask social fragmentation and homogeneity in the larger urban environment. The manufactured urbanity of the new developments presents a picturesque form of sprawl (Leung 1995): no more authentic than a gated enclave. Plastic columns and *faux* windows in Celebration, European columns in Poundbury, or anachronisms in other new urbanist projects fabricate a history that bears little relation to the places that claim it. Can front porches too narrow to accommodate chairs claim to revive the public realm?

Al-Hindi (2001) has said that new urban form generates prescriptive landscapes. Whose style do these places represent? Western Canadian developments present imported examples of New England brownstones; southern-style cottages appear in Ontario communities that face incongruously cold winters (see Figure 8.2). Rather than creating opportunities for redeveloping indigenous vernacular building types, new urbanism threatens to spark a new homogenization of tastes. On the one hand, as Duany (2002b:7) says, good copies are better than bad originals. On the

Figure 8.2 Whose history? Northern cottages
Clair Creek Village in Waterloo ON features cosy cottages that would seem appropriate on a sunny southern coast. The homes are clustered around a green inside a faux-gate entry.

other hand, post-war suburbs suffered from the homogenization induced by relentless copying of popular examples: do we want to replicate such standardization? What happens to diversity if we endlessly clone a few successful (sub)urban prototypes?

While developers have created 'town centres' and 'village squares' to accommodate the latest trendy retail outlets for affluent shoppers, the new urbanists have said little about the impact of the projects on old town centres and shopping malls. Perkins (2001) suggests that larger communities may suffer a net loss of vitality as downtown shopping districts decline or disappear through competition provided by the new 'town squares'. Those living near displaced shopping districts may then have to travel further to find retail outlets to meet their needs. Already threatened by big-box retail and regional malls, authentic old main streets may face yet another challenge to their viability from new commercial areas with the touches designers define as yielding 'true' urban form. To remain competitive, some older downtown areas have turned to new urbanist projects as well, replacing an earlier local character with the latest retro styles.

New urban developments in Asia and Europe seem less committed to traditional or classical urban forms. Designers in those contexts continue to experiment with contemporary forms and building materials, although generally at high densities and with a modern mix of uses. At the same time, though, new developments in those areas threaten the traditional blend of commercial and industrial uses in the city. Older retail precincts in many smaller cities in Japan are in rapid decline as affluent consumers drive to new shopping centres. Small-scale butchers and greengrocers in the UK face considerable competition from bigger grocery stores that cater to once-a-week shopping practices. Commercial trends are changing everywhere, and processes of standardization are well entrenched (Giddens 1991).

Are new urbanism projects generating communities that have the potential for long-term viability? To a large extent, that remains to be seen. Some will undoubtedly have long staying power. Some, though, seem likely to become demographic enclaves. Many new urban projects tend to attract small households with few children. Are the elderly and shop clerks moving into the accessory units, or do even the units designed to enhance diversity generally accommodate housekeepers, nannies, or the immediate family of the owners? Do these communities provide a place for everyone, or do they keep everyone in their place? New urban projects lack the diversity and messiness of real towns.

ARE THE DEVELOPMENTS DEMOCRATIC?

While modernism relied on the expertise of the designer to make decisions about the form of development, new urbanism generally seeks participatory involvement from the stakeholders in the development process. The advocates of new urban approaches use a range of participatory devices, but the design charrette is arguably the favourite. In the charrette, local officials, community residents, and designers work together to develop the vision for the new community. Over the course of several days, the project takes shape and consensus builds around the nature of the development.

Educating and engaging the public is an important component of new urbanism (Bohl 2000; Duany et al. 2000; Krier 1998). Proponents' faith in new urbanism to create more meaningful communities encourages them to spread the word far and wide. For instance, Krier (1991:119) says that traditional neighbourhood design 'will allow a much larger range of people and talents to become active citizens, in the full meaning of that phrase'. Good communities will become places that support good citizens, and good citizens will support new urbanism.

Design charrettes generate an amazing level of consensus on design options. Duany (2002b:7) says that 'Democracy leads inexorably to traditional architecture'. New urbanism uses the language of collaborative planning, and pushes for consensus-building and consultation. To a large extent, it also draws on the rhetoric and ideology of strategic planning, referring to participants as 'stakeholders' who should be engaged in the process at key points. At the same time as new urbanism welcomes a level of citizen participation, however, its fear of local opposition to projects is palpable, and its ability to accommodate diversity may be quite limited (Day 2003).

In the charrette process, the rhetoric of local control encounters the reality of slick graphics, romantic watercolours, and celebrity designers. Difficult policy or environmental issues are set aside as participants focus on design questions. Although the participants may see local concerns in the outcomes, an outside observer may also read professional values in the plans. Does the commonality generated in charrettes reflect underlying cultural values that designers can draw out of residents across diverse communities, or does it reveal the persuasive appeal of design professionals in the charrette process? With the wide media interest in photogenic new urbanist communities, we cannot easily separate fashion fad, consumer preferences, expert opinion, and democratic choice.

Marxists may see new urbanism and smart growth as the latest incarnations of the capitalist paradigm of urbanization. An economy that depends on continued growth for its survival needs new approaches when production lags. The choices offered by new urban forms, a Marxist analysis may suggest, are less about what the people want than they are about generating and sustaining economic activity in the market. Who benefits from projects like HOPE VI when market housing replaces public housing? The property owners around public housing developments reap considerable benefit from the reduction of concentrated poverty, but the disadvantaged residents of renewed places often find their communities disrupted and separated. Who benefits from mixed-use commercial projects like Melrose Arch in South Africa: arguably not the impoverished blacks who can never hope to shop in an upscale entertainment centre.

One of the truisms of contemporary life is that people resist change. Often this means that the strategies that facilitate democratic action do not tend to generate innovative responses as much as they reinforce the status quo. As Filion (1996:1654) says: 'Residents' empowerment thus serves to perpetuate forms of urbanization that were originally tied to Fordist consumption patterns.' Even in cases where new urban forms supplant modernist constructions, conventional patterns of authority and consumption survive the rhetoric of resistance.

Even the new urbanists have recognized the challenges of participatory decision making. Calthorpe and Fulton (2001) and Duany et al. (2000) argue that effective planning must occur at the regional scale, in part to escape local concerns and influence. Duany et al. (2000) are most explicit about this issue, noting that in Florida few initiatives to change regulations passed because governments feared to anger local residents.

> True participatory planning should not be confused with occasional orchestrated set pieces or legally mandated public hearings; rather, it should include community design workshops, citizen's advisory committees, constant media coverage, and an ongoing feedback process.
>
> That said, it is painful but necessary to acknowledge that the public process does not guarantee the best results.
>
> (Duany et al. 2000:226)

In other words, democracy may undermine design quality.

The new urbanist approach contrasts democratic participation with expert decision making by qualified designers who understand the principles of good community form. Duany et al. (2000:226) claim that because true representation is so often missing, decision makers must instead rely on principles; they then lament that principles are sometimes lost in democratic proceedings. Krier similarly finds democracy lacking (1991:119): 'The bad news is that democracy is capable of colossal environmental blunders.' The implication: it is easier to establish good principles of design than to secure a perfect democratic process. Principles must take primacy. Good design trumps democracy.

A detailed review of new urbanism texts suggests the commitment to democracy is weak. In describing the planning process for new urbanist projects, Duany et al. (2000:179) first explain the significance of the detailed master plan prepared by city governments. Such plans are prepared through a public process; once enacted, the plans guide development in prescriptive ways, setting the community vision. Communities may face a battle to get the plan in place, but in working through the process local governments create predictable environments that help them to avoid small battles over projects. After acknowledging the role of democratic processes in producing the plans, the authors then suggest that governments should enact master plans as quickly as possible to prevent the status quo from prevailing. Duany et al. (2000:180) proudly report that they set a record in Stuart, Florida, by having a plan presented to and adopted by decision makers within four hours. How much participation did community residents have in that four hours to review and discuss the plan? They claim that a 'fully open, interactive, public involvement process' generates successful plans (Duany et al. 2000:181) rings hollow.

The new urbanists want involvement in the vision process, but they often include converts and believers. The participants in charrettes seldom fully represent the community of interests in a project. While new urban designers advocate variety in form to accommodate urban diversity, they have difficulty tolerating opposing views about the form or character of the urban environment. Their reaction to critics and sceptics can prove quite hostile.

One of the features that accompanies new approaches to urbanism in Asia today is the opening of traditionally closed decision making processes to a level of public participation. In Japan, *machizukuri* ordinances give citizens new opportunities to influence planning outcomes in redevelopment areas (Koizumi 2004; Sorensen 2004). In South Korea as well, residents are seeking greater democracy in contemporary development processes (K. Kim 2004). While some countries continue to resist democratization, urban residents often press for a say in the future of their cities.

Many North American new urbanism projects are private communities managed first by development companies and later by home owners' associations.

European projects may also have extensive codes and restrictions that limit choice within new urban communities. While some may argue that home owners' associations are the ultimate in local democracy (since owners manage their own community), others argue that these organizations have the power to ignore democratic rights that people enjoy in public communities (McKenzie 1994). In seeking to control the public face of these projects, designers apply rules that limit people's choices about how to use and control the areas outside their homes.

Thus while we see evidence that some new urban approaches favour democratic processes, some forms of new urbanism resist the kind of intense and ongoing participation advocated within mainstream planning theory. Despite demands from community residents for the right to participate, and national governments that feel compelled to spread the message of 'freedom and democracy' to parts of the world that may have not have sought it, the new urbanists see a limited need for public involvement and private design choices in the well-designed community. The verdict on whether the developments are democratic is mixed at best.

ARE THE DEVELOPMENTS DIVERSE?

One of the key features of new urban projects is their propensity for mixing uses on a relatively fine scale. Duany *et al.* (2000) say that as long as a project is mixed, then it helps to avoid sprawl even as it accommodates growth. For the neo-traditionalists, then, the *pattern* of development is more significant than the location or amount of growth. The transit-oriented new urbanists who advocate regional approaches have a greater concern about location, suggesting that infill is preferable to greenfield development.

At the very least, the traditional neighbourhood design and urban village proponents believe that neighbourhoods need corner stores. Duany *et al.* (2000:187) say that corner stores will limit car trips and build community bonds better than social clubs can. The focus on providing local retail could lead one to conclude that the new urbanists see shopping as among the most important community-building activities. New urban developments may thus offer the appropriate domestic environments for a consumer society.

The rhetoric of new urban approaches consistently highlights the importance of diversity and mix. The reality of new urban forms, however, typically reveals limited diversity. Few projects have a substantial range of uses. Few have a comprehensive demographic profile. Robbins (2004) notes that new urbanist projects are rarely complete towns; they seldom include appropriate local retail. Many do not have adequate transportation options. Moreover, most of the communities are populated by wealthy, white professionals: double-income, no kids.

Frantz and Collins (1999) lament the lack of diversity in Celebration, despite the good intentions of designers to create a complete community. Only a handful of black residents moved to the town. For its first several years, Celebration had no hardware store or hair salon, only a selection of shops that catered to tourists visiting America's new perfect community. As I walked around neo-traditional developments in Maryland in early 2004, many women of colour pushed white babies around in carriages on a lovely spring day. In these communities of $700,000 homes, visible evidence of diversity appears to mask a class and ethnic structure of extreme inequality and separation.

An increasingly common critique of new urban developments discusses the normative and moral position implicit in a movement that promotes the 'traditional'. Does a backward-looking style welcome everyone? Al-Hindi (2001) sees neo-traditional developments as reflecting growing social and political conservatism. These communities rarely provide homes for poor people, immigrants, or people of colour. Do any include group homes, women's shelters, or half-way houses? Dowling (1998) argues that architectural controls constitute a code word for homogeneity. The designs stipulated represent a way of marking class and identity. They celebrate a history of colonialism and control.

Janet Smith (2002) notes that the focus in America is on mixing housing in formerly poor neighbourhoods: she sees little interest in changing affluent areas to integrate the poor. The reality of the contemporary city reflects the fear of concentrating poverty. Breaking up concentrations of poor households by distributing them around in a one-in-ten proportion, as suggested by Duany, may make poverty less visible, but it may not serve the integrative objectives envisioned by the proponents of the strategy.

The mixing promoted in new urban projects is limited (Grant 2002b). Usually designers propose to mix residential, commercial, and office developments. Rarely do they want industry: more often they are replacing industry as it abandons the city. When they develop commercial components for projects that are primarily residential, they may have a hard time finding tenants for the commercial outlets (Thompson-Fawcett 2000). In a period when chain stores on arterials are displacing small convenience stores in established neighbourhoods, attracting neighbourhood retail to new projects remains a significant challenge.

While traditionally diverse in use, East Asian cities have limited ethnic variation. In contemporary developments, class separation is increasing as new suburbs are built. New residential areas prove relatively homogeneous in cost and style, and hence may attract similar households. In suburban areas, residential areas include some mix of uses, but with less commercial than in the older cities. New suburban projects are more frequently car-oriented, with residents driving to shopping centres for their retail needs.

Hence we find that while theory in all kinds of new urban approaches in almost every region promote accommodating diversity and mix, the overall development trends more often generate segregation, except where governments intervene to provide affordable housing or where well-established mixes persist. Even though diversity is promoted in the popular culture, and the systemic barriers to it may be removed, the market functions to restrict diversity through economic barriers that limit business choices, or by codes and covenants that control property use and design. Despite design strategies that promote diversity, practice resists difference.

ARE THE DEVELOPMENTS EQUITABLE?

The European urbanists are concerned to provide a level of social equity in their projects. Several include financing high proportions of social housing in the mix of new projects, supported by government or charitable funds. Given that most of the projects are apartment units, the products are likely the most affordable possible. This enhances the prospects of greater equity through development.

Concerns about equity seem less commonly articulated amongst the developers of new projects in East Asia. Housing costs are high because land costs are high. Consequently, apartments are the most common form of new housing offered. Governments and employers provide most affordable housing, often in large projects with good access to public transportation. By contrast with the dismal state of public housing apartment projects in North America and the UK, projects in Asia are often well-maintained and well-located. A desire to integrate the poor to improve their life opportunities is less often expressed through redevelopment projects.

New urbanism is often presented as a strategy for ending the economic exclusiveness of the suburbs and generating more equitable communities to integrate the poor into the wider urban fabric. Ending segregated uses contributes to developing more equitable cities. Concentrations of privilege and poverty have to disappear. In the American context, though, promoting equity does not mean building large areas of affordable housing. Bohl (2000:779) says that: 'One reason for the growing application of New Urbanism in inner cities is to support policies aimed at reducing high concentrations of poverty by creating more diverse, mixed-income neighborhoods.' As we have seen, Duany says that cities need more middle class people instead of affordable units: distribute affordable housing as sparsely as possible to ensure integration (Duany et al. 2000). Thus not only do new urbanists advocate using design and quality architecture to reduce stigma for poorer households, they envision their developments as social projects that allow the poor to assimilate to the culture of the middle classes. New urbanism is a civilizing project.

The critics argue that American urbanists have little concern for equity. For instance, the editor of an issue of *Planners Network* bulletin wrote:

> While the 'movement' has gained plenty of attention in the popular media, we remain concerned that the movement's focus on the built environment

> masks deeper issues of social equity and power. The simplicity of new urban prescriptions for built form seems to run against the complexity of urban problems.
> (Milgrom 2002:2)

By removing its outward manifestations, new urbanism would render poverty invisible. Claims that neighbourhoods can absorb a one-in-ten admixture of poor households are made purely on a hypothetical basis, since the new urbanists offer no evidence to support their arguments about integration and equity enhancement. Neither do they make clear what should happen with other poor households which cannot be accommodated because they may exceed 10 per cent of the population. Does the good community set a limit on how many poor people it will accommodate?

Duany often argues that gentrification is good for communities (Duany *et al.* 2000). Gentrification represents principally a sign of revival and hope for neighbourhoods, not a loss. New urbanists thus urge planners to encourage gentrification, while they may dismiss those who criticize gentrification by saying that the complainers merely hope to maintain a failing power base. This focus on the visual aspects of the built environment and the improvement in property values may make new urbanists seem insensitive to the social consequences of displacement and community disruption. Growing homelessness is a reality of the post-modern city, yet it rarely merits attention from new urbanists. Instead, the focus in prescriptions for change is on bringing tax-paying residents back to the city (Duany *et al.* 2000:173).

Are the environments in well-integrated new urbanist projects or urban villages that accommodate both affluent households and poor families equitable? Those purchasing homes in the developments have a choice about accepting restrictive covenants that limit their options in decorating their homes or restrict their behaviour on their grounds. Those moving into public housing have no such choice (Marcuse 2000; Thompson-Fawcett 1998). They find themselves bound by paternalistic judgements about good taste in the built environment and socially-appropriate behaviour in the laneways. As Thompson-Fawcett (2003a) notes with regard to Poundbury, poorer households in these communities rarely interact with their affluent neighbours. Rather than promoting equity and sociability through propinquity, such projects remind the poor of their station in society.

The focus on the public realm reveals new urbanism's concern about providing equitable and accessible public spaces. Designers regularly talk about the street, the square, the green as formal elements of urban structure to create open environments for civic activity. The rhetoric contrasts, though, with the reality that many of the projects create private spaces. Commercial spaces and private roads provide opportunities for exclusion. The facilities and amenities within many of these communities are 'club goods' (Webster 2002) available only to residents and their guests. Webster and Lai (2003) ask, are these places about optimal inclusion or optimal exclusion? Given the iconography of affluence in the buildings and civic spaces, marking status seems an important element of new urbanist community building.

ARE THE DEVELOPMENTS SUSTAINABLE?

Post-war development practices clearly were not sustainable: they consumed land and resources at rates impossible to replenish, and far beyond the amount reasonably required for good living conditions. Do new urban approaches offer development options that prove more sustainable?

New urbanism seeks to be comprehensive in its outlook, beginning at the building and proceeding up through neighbourhoods to districts to generate sustainable regions (Calthorpe and Fulton 2001; Katz 1994). It attempts to be holistic, encompassing, integrative, to account for environmental as well as economic and social factors. Like the modernists before them, the new urbanists hope to find integrating solutions that may solve multiple problems and create better places for living (Beauregard 2002).

Gordon and Tamminga (2002) demonstrate that open space plans for new urbanist projects may conserve ecologically sensitive features. The plan for Markham identified environmentally sensitive areas. An evaluation after development according to neo-traditional principles showed that the builders protected most of the environmentally special features of the site. But not all projects reflect the same level of environmental responsibility. For instance, Frantz and Collins (1999) criticize the approach to nature taken in Celebration. They suggest that the desire to squeeze out extra building lots led the company to drain wetlands, build lakes, and cut large trees. Similar complaints about poor environmental protection have been levelled at other Florida new urbanist projects (Audirac et al. 1990). Zimmerman (2001) questions what conceptions of 'nature' fringe developments like Prairie Crossing in Illinois actually seek to 'sustain'.

Much of the new urbanist development has occurred on greenfield sites, transforming farm land, desert, or forests to housing. Such suburban expansion leaves new urbanist projects vulnerable to criticism that they cannot contribute to sustainability. Urban infill and redevelopment have been increasing in recent years, however, adding to the claim that new urban approaches can provide a sustainable development option.

Brown and Cropper (2001) note that houses in a new urbanist project in Utah were larger than units in a comparable conventional suburb, but on smaller lots. They describe the average house at over 300 sq m with lots more than twice that. Each year the average new home built in North America is larger than the average of the previous year. Even as household size declines, living space per unit increases; unit densities increase, but urban population densities may drop. Is continued expansion in house size sustainable? It certainly results in increasing demands for building materials, and generates additional demands for energy to service the residents. Even if the lot size diminishes, we see little indication that overall material and energy requirements will be reduced.

New urbanists generally argue that smaller lot sizes contribute to sustainability. Brown and Cropper (2001:405) say that: 'In a suburban context, the question of ecological sustainability often hinges on whether residents will accept smaller lot sizes.' Can smaller lots make enough of a difference? They may reduce overall consumption of land, but ecological sustainability depends on many factors. For instance, how do residents use the lots? Covered with houses, garages, or pavement, even small lots may harm landscape function (Grant et al. 1996). Small lots limit residents' ability to grow their own food locally. Requirements to maintain an urban character in compact communities may restrict the potential for naturalizing lots to provide wildlife habitat. Expansion of settlement in areas where air conditioning is a pre-requisite of contemporary middle class urban living, as is the case for most new urbanist projects in North America, increases demand for fossil-fuel energy use, even if such projects could be successful in reducing car use. Many new urbanists seem to see sustainability as essentially a design problem. Their understanding of ecological issues and options for sustainable development seems shallow or unidimensional.

The latest twist to new urbanism in North America links it to smart growth. Smart growth promotes the design options of new urbanism, but adds government policy and incentives to accommodate and encourage growth. If the population keeps growing, as the smart growth advocates insist it must, and materials consumed per person keep increasing as they seem to do, then how can we hope for sustainable cities? At some point, we will exhaust non-renewable resources. Even renewable resources, like ground and surface water, may be threatened by expansion in the regions where new urbanism thrives. Using less land per building is only one measure of sustainability. Most of the other indicators of sustainability are not good. For instance, car use is not declining (Williams, Burton and Jenks 2000), nor is energy consumption per capita diminishing. The most popular vehicles in North America are large gas-guzzling vans, SUVs, and trucks. Growth may be necessary to our capitalist economy, but it cannot continue indefinitely. Suggesting that new urbanism can promote 'sustainable growth' (Garde 2004) strains credulity. New urbanism is linked to an ideology that believes that design and policy solutions can bridge the inevitable tension between capitalist expansion and ecosystem limits. While the rhetoric proves inspiring, the reality may be less convincing. Our society is fundamentally unsustainable, and new urban approaches do little to alter that truth.

Ecological theory inspires new urbanists. In the absence of alternative theory to substantiate their claims, they find in ecological theory a language that resonates with a public concerned about the environment and searching for sustainable futures for our children. There are limitations to the use of ecological theory in this context, however, especially as it represents the application of an organic analogy rather than a full theory of urban form (Thompson-Fawcett 1998). Like evolutionary theory before it, ecological theory is often inappropriately applied. While the city constitutes human habitat in the broadest sense, human constructions do not operate solely by the principles that govern natural systems. Human decisions and actions can rapidly alter the landscape through technologies and management systems that have no counterparts in ecosystems. Our cities are artifacts more than they are organisms: the products of countless human choices through time.

While the new urbanists often use the rhetoric of ecology to legitimate solutions, they do not incorporate all of its principles into their theory. For instance, they do not recognize that ecosystems adapt continually to change, and never remain in stasis. As Durack (2001:68) says, 'life is not an equilibrium condition'. Ecosystem niches are not codified for species composition: they are fluid, dynamic, and responsive. Yet the rules that new urbanists apply to preserve their urban niches restrict adaptive response. We cannot know what is sustainable over the long term (Durack 2001): the best options may not become apparent for another millennium, when our descendants see what remains standing (Grant 2004). As Malcolm Smith (2004:online) notes: 'The last thing a healthy environment wants is to be frozen in the artificiality of new urbanism, like many town centres are now locked into ideas of the 1960s.' Finding sustainable choices requires us to maintain maximum flexibility and adaptability. The rigid codes of contemporary new urbanism may limit future options. To ensure that we can provide good communities that meet the needs of next generation as well as our own, we need an open urbanism with constant review and the potential for adaptive response.

POWER IN NEW URBANISM

While new urban approaches have generated beautiful new urban districts in many places, I have suggested that they appear less successful in attaining other attributes important in the good community. In this section, I argue that what Lynch (1981) calls 'hidden values' are implicit in new urbanist practice. Hidden values, says Lynch, are rarely articulated in theory but often achieved in practice. For planning he suggests these may include maintaining political control and prestige, disseminating an 'advanced' culture, dominating a region or people, removing unwanted people or activities, and making profits. In sum, planning involves the application of power.

The new urbanists promote diversity, affordability, and equity in their books and articles, yet our analysis of the practice of new urbanism reveals a hidden power dynamic. New urbanists build communities that essentially serve the needs of an urban élite and the segment of the economy that depends on urban development. In some ways, then, new urbanism reproduces and reinforces existing power structures. Its claims of equity and empowerment remain unrealized: moreover, they may offer little more than rhetoric that masks practices that increase disparity.

In the contemporary development market, new urbanism has a certain display value. This not only gets it on the cover of magazines and featured in news reports, but also attracts a clientele eager to display their relative affluence, worldliness, and success. Creating a distinct built environment confers status on those involved: it allows the new intellectual/information age élites to display their good taste. For the residents, design codes may also provide a means of demonstrating political and social values.[1] Those selling and inhabiting these townscapes have appropriated the language of other theories to label their communities as sustainable, smart, healthy, and environmentally responsible. Perhaps residents persuade themselves that they can consume this housing product without guilt. John and Jane Q. Public need to believe that they are doing the right thing in buying a large, expensive loft, a replica brownstone, or a modern Victorian cottage: they have invested in the good community (see Figure 8.3).

Franz and Collins (1999) and Ross (1999) describe the residents of Celebration as self-selected for that community. People moved to the Disney new town because they liked the look and the intent of the place. They wanted to participate in a great urban experiment, and they paid more to do so. Like other new urbanist communities, Celebration's builders and designers projected a myth of diversity and inclusion, but the community as built proves relatively homogeneous. Nonetheless, residence in the town conveys a particular status on the occupants.

New urbanism employs an image of social order associated with the model of small town, urban quarter, or urban village: one of privilege and hierarchy. The models

Figure 8.3 Fantasy Village
New urban approaches are hardly sustainable or smart if they undermine landscape function, consume excessive materials and resources, and ignore the needs of the less affluent. The dream that retro styles can recapture the good community proves a fantasy.

replicated date to a time when the owners employed domestic servants who often lived under the attic dormers or in flats over the garages. They recreate the image of a landscape of female inequality, a dominant nuclear family unit, racial and class segregation, and limited immigration.

> The Urban Villages Group and Congress for the New Urbanism, for example, are in many ways explicit about their aims for an urbanity which encourages civilized community and family relations that contrast with the apparent loss of community (patriarchal?) control in modern cities.
>
> (Thompson-Fawcett 1998:189–90)

Clearly the designers do not intend to reproduce that hierarchical social order, yet they fail even to comment on the possibility that the built environment conveys messages of power. Frantz and Collins (1999) note with irony that the pavilions at Celebration look like replicas of the slave markets of St Augustine, yet the designers wonder why few black families moved to town. If, as Upton (2002) suggests, planning is a form of 'spatial ethics', then we must consider the hidden values implicit in our design guidelines and planning regulations. Since the Victorians preserved their challenged aristocratic values in parks, museums, and galleries (Kellett 1969), and the late nineteenth-century industrial barons of America protected their élite values by funding operas, ballet companies, and symphonies (Levine 1988), perhaps we might suggest that late twentieth-century élites hoped to conserve their values through private streets and architectural codes and covenants. As Douglas and Isherwood (1996:66) note, 'the rich who call the tune are continually changing it, too'.

Many new urbanist projects and organizations have been built primarily because of the patronage of powerful men. The Prince of Wales constituted the driving force behind Poundbury and many of the organizations promoting new urban forms in the UK and Europe. Galen Weston funded new urbanist projects in Florida and facilitated development of the Congress for the New Urbanism (Thompson-Fawcett 2003b). As the grand urban projects of earlier generations depended on men of power, so do cutting-edge contemporary developments. Versailles came about because of the audacity of Louis XIV; nineteenth-century Paris revealed the power of Napoleon III; Rome expresses the legacy of popes commanding the fealty of the masses. How should we interpret the new urban forms with their messages of democratic engagement when their proponents curry the favour of the most powerful élites in society? Important men[2] have the potential to advocate the right choices to generate good communities, yet critics will always find reasons to suspect their motives.[3]

Certainly new urbanism reflects a nostalgia for the conformity of the small town, ethnic quarter, or rural village where everyone knew each other and behaved responsibly. It reflects a belief in the civilizing influence of good design: power embodied in the built form. As Banai (1996:184) says, 'the neotraditional town can be seen as a (piecemeal) spatial strategy to create some stability by means of historicized symbolic landmarks to fix the suburbs'. The design reflects an overall desire to control the behaviour of people in space, and the belief that bad behaviour results from unattractive and inappropriate form. Traditional forms return with the hope that they might contribute toward reimposing historic social controls.

While many western designers and planners describe the urban environment in Japan as ugly and even chaotic, social behaviour there has remained quite controlled and conformist (Shelton 1999; Sorensen 2002). Urban design in Japan is functional, but seldom aesthetically pleasing to western designers. The lack of adherence to classical design principles does *not* coincide with weak social control, as western designers seem to assume in discussing the problems of North American cities and suburbs. Strong cultural controls on behaviour have kept cities orderly and civilized in Japan: crime rates remain low. Consequently, new urban approaches in Japan do not expect design to generate social control. An architectural régime of free expression and rampant modernism can coincide with a culture of good behaviour. A thoroughly modern city can be a 'good community' in many ways.

Social controls that once conditioned behaviour have disappeared in the West since the 1970s (Lianos 2003), replaced by an 'anything goes' culture and a mass media full of bad behaviour disguised as entertainment. 'Gangsta' rappers and mass murderers become media celebrities; professional athletes arrange hits on their enemies, agents, or relatives. Some see a civilization in decline. Designers note a correlation between declining civility and suburbanization and tend to assume causality. The suburbs encourage bad behaviour, they say. New urbanists blame a range of ills on the anomie and lethargy of the suburbs: youth violence (Kunstler 1999), road rage and teen suicides (Duany et al. 2000:62, 121), and, the latest target, obesity (Arendt 2001; Kreyling 2001). If we look at wider cultural trends that provide children with violent movies and video games as entertainment, and that encourage consumers to treat themselves to super-sized meals on a daily basis, then we might reasonably conclude that we find no shortage of factors to blame for our modern dysfunctions. Instead, the new urbanists see the worst elements of bad behaviour in contemporary society as caused by sprawl. In the American design context, we seem unwilling to look beyond the built form to identify other features of contemporary culture as undermining civility and good behaviour. As the critics of urban blight in mid-twentieth century advocated radical surgery to remove the cancer of decay in inner cities, the new urbanists at century's end argued that new urban forms could cure the ills of the suburbs and of the remaining pockets of poverty in the city.

The new urbanists advocate an urban code to shape building volume, articulation, relation to street, and building type. As Duany et al. (2000) explain, the code describes a proactive vision and shapes public space to achieve it. The new urbanists also argue that good codes can ensure that environments support positive behaviour. 'Here it is the place of the code to keep the city civilized, exactly as laws are intended to do,' Scully (1991:20) wrote about Seaside. Since laws and authorities fail to manage civility effectively, then communities must expand their measures of control. Codes control form 'to establish and guarantee the value of the home that has been purchased' (Robbins 2004:223). Moreover, they define the good community and resist the problems of the city by setting rules about the character of place. The designer frames rules that operate in perpetuity to retain the vision of the founders.

One may see the design codes as operating not only to normalize space, but to control those who occupy the space. The residents of private communities consent to control both their built environment and their behaviour within the space through conforming to the codes (Lianos 2003). As governments move to extend their regulations to include design elements, control expands to wider spaces. In a Foucauldian sense, the new urban form plays a disciplinary role, increasing surveillance and territorial control to induce conformity and compliance (Foucault 1977). As Lynch (1981) recognized, urban forms like gates and boundaries, but also public features like monumental axes or bilateral symmetry, can convey implicit messages of power. Design disciplines choice and behaviour in the controlled city. The judges of normality are everywhere (Foucault 1977), from the 'porch police' in Celebration (Frantz and Collins 1999) to the urban design inspectors working for local governments.

The new urbanists support Jane Jacobs's (1961) call for eyes on the street. The new urbanists implement her ideas by ensuring that windows overlook public spaces. In the contemporary city, we are obsessed with visibility (Koskela 2000). Good public space by definition provides opportunities for full scrutiny.

> He who is subjected to a field of visibility, and who knows it, assumes responsibility for the constraints of power; he makes them play spontaneously upon himself; he inscribes in himself the power relation in which he simultaneously plays both roles; he becomes the principle [sic] of his own subjection.
>
> (Foucault 1977:202–203)

For the new urbanists, the good community applies public scrutiny everywhere outside the private context of the home. New urbanism seeks to operationalize the panopticon as Foucault explains it, creating spaces in which everyone believes that someone may be watching, and adjusts behaviour accordingly.

Thompson-Fawcett (1998) asks whether the design codes represent a form of cultural imperialism: set by the designer and producer to enforce particular values fixed in time and place. We might argue that the interests of capital generate the values that dominate the landscape in this context. Alternative tastes are coded out. '[S]ocial interactions increasingly operate in socio-technical environments within which negotiation is not possible', says Wood (2003:237-8). No point in trying to convince your neighbours that you should be able to paint your house purple if the rules say you cannot. In a free market, consumers may tie themselves to contracts that limit their design options, and those of subsequent purchasers. McKenzie (1994) has documented the proliferation of private communities in the USA; millions of Americans live in enclaves subject to extensive codes, covenants, and restrictions. The same is increasingly true in other countries. Consumers willingly yield their rights to choose. But, as Robert Stern, one of the planners of Celebration, is quoted as saying: 'In a free-wheeling Capitalist society you need controls – you can't have community without them ... I'm convinced these controls are actually liberating to people. It makes them feel their investment is safe. Regimentation can release you' (cited in Robbins 2004:223). Surrendering to authority frees consumers to protect their investments. Gross (1980) calls this 'friendly fascism': the black boots are missing, but power is no less persuasive.

The design codes operating in private communities arguably represent a particular class orientation. For instance, the regulations at Celebration mandate window coverings that appear white from the outside, prescribe front porch furnishings, and limit choices of planting materials (Frantz and Collins 1999). At Poundbury, initial purchasers choose what colour to paint their front doors (from the limited palate offered), but subsequent residents may not deviate from that option (Hardy 2003; Thompson-Fawcett 2003a). Most private communities, whether new urbanist or otherwise, do not allow residents to hang laundry out to dry. No chance that students will hang flags to cover windows, that wildflower meadows will sprout on manicured promenades, or that residents will fix broken cars in their yards. The rules reflect an upper-class aesthetic of the house as display space. They convey power relations through buildings (Koskela 2000) and the grounds that surround them.

At the American Planning Association conference in Washington DC in April 2004, many sessions advocated 'form-based codes'. APA is encouraging municipal governments to consider new zoning regulations that enforce particular design standards in urban areas. These codes may replace or augment existing zoning regulations. Thus the designer's aesthetic becomes embodied within a publically-imposed framework of rules and obligations. Whereas conventional regulations focus on land use, form-based codes set design regulations such as build-to lines or building-height-to-street-width ratios. Sitowski (2004) wonders whether such codes will be challenged in law. Is it fair and reasonable for planners to legislate the pitch of roofs or the proportions of windows? Certainly planners can argue that controlling activities in space has legitimacy, since use can affect health, safety, and efficiency

(Myers 2000). Aesthetic standards, however, are far from universal. Does the state exceed its legitimate police powers in seeking to regulate taste?[4]

In the post-modern city, we search for meaning in the past. Some would argue that the new urbanists' search for tradition serves to exclude many in our cities. Typically, the heritage celebrated in new urban forms is that of the winning colonial or imperial powers. While we talk about diversity and inclusion, we tend to ignore elements of our social history that represent women, aboriginal peoples, or minorities (Hayden 1995; Veregge 1997). 'The characteristics that Duany and others cite as useful precedent have been extracted from historical time and divorced from the processes of change that shaped them', notes Veregge (1997:51). They are ahistorical, selected for particular aesthetic virtues. The experts use a narrow definition of tradition, side-lining alternative histories that we might choose to reconstruct and celebrate. Theirs is a reductive science, trying to identify correct types in order to replicate them (Turan 2004). By contrast, some would argue, we could explore varied histories and insist on the relevance of context in selecting appropriate inclusive choices.

As Veregge (1997:59–60) notes:

> In its exclusion of temporal change and cultural diversity, 'the traditional American town' model implicitly speaks of New Urbanism's basis as a response to widely held fears current among the property-owning class about neighborhood deterioration and the growing presence of populations of non-European origin.

New urbanist forms, especially in the North American context, help to construct social distance in societies that are becoming increasingly diverse through immigration. The value of diversity promoted in the theory stands in stark contrast to the message of homogeneity portrayed through the developments. Most of the inhabitants of American new urbanist communities are affluent, white Anglos.[5] In reconstructing New England brownstones or southern colonial mansions they send a message of exclusivity. They actively resist difference.

New urbanism seeks to define social problems as technical ones (see, e.g., Krier 1998; Dumreicher *et al.* 2000). For example, Krier (1984a:46) explains that

> High density and increased exploitation of urban ground have been wrongly identified as being responsible for the inhuman condition of the nineteenth-century city. Instead the badly lit light-wells, the polluted streets and the endless corridorial spaces inside the vast blocks were in fact the result of a *wrong typological choice: the large urban block.*

A technical criticism of the typological choice could have resolved the problems, leading to better urban form, he says; instead, a more general critique of the industrial city led to a new model of development, with its subsequent problems.

Those who accept that the problems of the city are technical – that is, embedded in form – may then argue that only the designer provides the expertise to solve the problems through good design. This understanding clearly privileges the architect/designer/planner. A strong streak of authoritarianism runs through the new urbanism with its demand for the designer to assert control: D. Hall (1998:34) calls this 'design totalitarianism'. The master planner/designer manages the process, articulates the vision, and sets the codes that will obtain for all time.

Krier (1998) makes these arguments quite forcefully. He says that communities should commission an independent designer with vision and authority to make the master plan to set the rules. He argues that the public demands input primarily because design has failed to provide satisfactory outcomes in the past. If communities had good design, the implication is, people would not clamour to participate. Democracy should not mean chaos and disorder, Krier says: harmony is at the root of civilization. While public interaction and participation are required, 'the architect will be located in a more critical position than conventional practice affords' as the new community order is articulated, write Dumreicher *et al.* (2000: 293). A designer with the authority to order space can restore good communities.

People need education from designers because cultural democracy is so poorly developed that ordinary folk do not understand good design, suggest some new urbanists. 'As homebuyers ... become familiar with these alternatives and become better educated about what kinds of options, advantages, and trade-offs might be available, their opinions of new urbanism design may change' (T. Alexander 2000:46). Designers have faith in the power of education to shift popular values closer to professional values. New urbanists show less interest in the idea of participatory design that involves citizens controlling the design process (Fung 2001).

Who benefits from the new urban strategies? Attention on urban design has created new job opportunities for architects, planners, and urban designers. A few designers like Krier, Duany, and Calthorpe have become famous, influential, and wealthy through promoting new urbanism. Professions eclipsed by engineers and accountants in the 1960s through 1980s have made a come-back. Strategies for redeveloping inner city precincts have improved land values in the cities, driving revitalization projects that benefit property owners. Projects that facilitate higher density development enhance land values for suburban growth. 'In the short term, [new urbanism] appears primarily to benefit design professionals, developers, and the predominantly white, upper-class homebuyers who choose to live in its communities', claims Al-Hindi (2001:204). Whether wider constituencies will benefit over the longer term remains unproven.

Public and Private Interests

The codes imposed in covenants on deeds on properties sold in the early twenty-first century control management and prevent future owners from changing the elements central to approved plans. They inscribe the contemporary designer's vision of the good community on generations to come. Not only do they fix design in one place and time, they set a long-term understanding of what constitutes the public interest or common good. Place making today contributes to social development tomorrow and to long-term public welfare (Talen 2002). Thus new urbanism continues the legacy of adherence in the notion of the public interest as a legitimation for planning decisions. Critiques that hold that unitary views of the public interest deny a reality of conflicting individual and group interests or the possibility of change over time do not dissuade the new urbanists. As Talen (2001:online) says: 'I think that the correct response to this view [that there may be no common good] is that it is the dormancy of normative principles that is dangerous and problematic.' Contemporary values will be enforced over a long term, with the understanding that the designer has the best interests of the community in mind. In a sense, then, the new urbanists differ little from the modernists in their faith in the supreme authority of the expert (Al-Hindi and Staddon 1997) or in their own moral positions (Beauregard 2002), but they have trumped the modernists in employing legal mechanisms to protect their visions of the public interest over a longer time frame.

Faith in growth, a key premise of modernity, remains implicit in new urban planning. New urbanists often say that growth is inevitable. They think it can be sustainable. They have aligned themselves closely with the smart growth movement. Attractive residential and commercial areas make growth seem more palatable and acceptable to those who otherwise fear change. Vibrant urbanism has become the code for dynamic capitalism in the city. The powerful who benefit from economic expansion find new urban approaches offer useful marketing strategies for urban growth.

As Myers (2000:online) says: 'Planners want to be visionaries and preachers, not researchers.' Practitioners have little patience for research to figure out what works. They may exaggerate their claims, looking for strategies to derive support for the options they favour. Salesmanship and showmanship may divert attention from the critical issues at hand. Crane (2000) notes that more research to test the claims of new urbanism is essential. We need to understand the economic, social, and political implications of contemporary models of urbanism to know whether practice can bear out the rhetoric of choice, diversity, and sustainability. The real contribution of new urban approaches to planning theory and thinking are currently being tested in practice, but the results are not all in.

The key issues of mainstream planning theory in the late twentieth and early twenty-first centuries are arguably related to power and space. New urbanism pays little attention to these themes. Its proponents argue that new urbanism promotes democracy and participation, yet we see little evidence of the collaborative planning that planning theory suggests would comprise the major mechanism to implement democratic planning. New urbanists rarely refer to key planning theorists like John Friedmann, John Forester, or Patsy Healey, instead looking to architects for theoretical insights. New urban spatial concerns generally seem much narrower than the questions that interest planners and geographers (e.g., Cooke 1983; Knox 1993; Soja 1989). While some new urbanist authors have attempted to articulate a regional planning approach, we see that the theoretical development ignores the implications of globalization in favour of a focus on design questions. In sum, theory in new urbanism proves relatively weak and disconnected from theoretical developments in urban planning. This limits its ability to contribute to the discipline and renders it vulnerable to continued criticism within the planning academy.

CHAPTER 9

THE FATE OF NEW URBANISM

COMPETING URBANISMS

New urbanism has become an influential force in planning practice, especially in North America, yet it has made relatively little impact on international planning theory. New urban approaches are important rhetorically in many regions, extensively affecting planning discourse in Europe, Australia, and South Africa, for instance. Yet the effects on mainstream development practices – at least in the countries we have considered here – are more limited. Indeed, we might predict that the fate of new urbanism will likely parallel that of the garden city. In North America, for several decades it will shape physical form in a reduced and simplified way, gradually bereft of its mission and vision. As J.J. Stevenson said in a lecture in 1874 about the Queen Anne style (which had recently replaced Gothic revival at the time), 'the movement has become fashionable, and consequently in danger of becoming vulgarised' (quoted in Rubinstein 1974:60). *Plus que ça change*. New urbanism's original principles are already being watered down in practice as developers 'cherry-pick' the saleable parts of the concept. North American New Urbanism has been transformed to a broader and more marketable concept of generic new urbanism in application, as urban village models and smart growth notions set aside some of the initial principles. In Europe and Asia, new urbanism may well become lost amidst the many urban traditions that preceded it. In much of the developing world I suspect that new urbanism will find itself unable to compete with gated communities in appealing to new urban élites.

As planning theorists continue to work towards developing theory of good city form, some of the ideas of contemporary new urban approaches may well remain influential, while others become historical artifacts. New urbanism may have some longevity, primarily because it provides a powerful marketing device for developers competing to sell homes. Ultimately, though, history may well reveal the weakness of new urbanist approaches: namely, the commitment to urban form rather than to social or political reform as the means to create good communities. Practice will rapidly reveal its failure to achieve its ideals.

Since contemporary mainstream planning theory focuses on issues of power, space, and communication, new urbanism has not proven as powerful a theoretical construct as its advocates may hope. New urban approaches have drawn on theory eclectically, seeking support for pre-determined positions in an opportunistic rather than a systematic manner. This has left it open to empirical challenge and epistemological critique. While new urbanism resonates with those embracing a return to urban values, other key development trends in suburban form reflect and perpetuate radically different values, and hence undermine the long-term potential of new urban theory and practice.

New urbanism articulates a distinct model of the good community as a beautiful and identifiable place. In that, though, it constitutes part of a wider process in development whereby a variety of participants make similar claims about having mastered the formula to create good communities. In times past, people may have taken community for granted, but in the contemporary context, selling community has become key to moving property (see Table 9.1): this is true in conventional developments (like Portland Hills in Nova Scotia), gated enclaves (like SongBrook in Oregon), or new urbanist new towns (like Celebration in Florida).

THE SUCCESSES AND FAILURES OF NEW URBANISM

Krieger (1991:14) has called the proponents of new urban approaches 'Pragmatists with a vision'. It is an apt expression. Until the early twenty-first century, the new urbanists made little pretence of developing theory. They focused on building projects, thus spreading the philosophy of a new approach to dealing with cities through explicit actions. The success of the movement reflects its ability to align its design strategies with government development objectives and market interests. New urbanism offered an appealing vision to governments looking for ways to promote growth and urban revitalization in ways that could reduce the negative externalities associated with urban development. Municipal, provincial, state, and national governments in many countries have adopted some of its principles as official policy. With success in getting projects started and planners interested in the approaches have come initiatives to have new urbanism taken seriously in the academy: hence the need for theory to substantiate the messages of practice. The development of theory can also begin to put new urbanism practice in its context, deriving the lessons of its successes and failures.

Few observers would say that new urbanism has been unequivocally successful. It has not stopped sprawl, even though it has made development more attractive. It may exacerbate problems of affordability and inequity, especially in areas where governments hesitate to intervene to support or require low-cost housing. By condensing the built form, it may conserve land and resources outside the city, while intensifying local environmental impacts.

What lessons have we learned to date about the fate of the new urban movements? Does new urbanism meet its goals? The movement has achieved pockets of

Table 9.1 Selling community

SongBrook, Eugene, OR	Portland Hills, Halifax Regional Municipality, NS	Celebration, Osceola County, FL
gated community	conventional project	new urbanism project
At SongBrook, we realize how blessed we are. We are amazed at the wealth gathered in our manufactured home community. The wealth of friendship, caring, wisdom, experience, taking that extra step, going the extra mile, reaching out with a hand. A wealth of trust and faith that each one brings when they make the decision to become a part of SongBrook. For SongBrook is you, our friends, and there's no other place like it. Our community – our Family – is like no other. At SongBrook, we truly live community; we walk hand in hand.	Our master-planned communities are built with the individuality of our customers and the prosperity of the community in mind. They bear the hallmark of over 40 years of experience. The carefully designed and welcoming streetscapes are combined with the natural beauty of the surrounding environment. Our protective covenants and architectural controls ensure that your neighborhood will develop and grow as it should – as a warm and pleasant community that will have enduring beauty and enduring value.	Take the best ideas from the most successful towns of yesterday and the technology of the new millennium, and synthesize them into a close-knit community that meets the needs of today's families . . . A place where memories of a lifetime are made, it's more than a home; it's a community rich with old-fashioned appeal and an eye on the future. Homes are a blend of traditional southeastern exteriors with welcoming front porches and interiors that enhance today's lifestyles.
www.songbrook.com/page/page/733691.htm	http://www.claytondev.com/ResSection.ASP?item_id=257	www.celebrationfl.com

All web sites accessed 6 July 2004

density and mix in several successful projects. We can point to many examples of high-quality built environments quite likely to stand the test of time. To date, though, these projects seem relatively few and far between. While many towns constructed were built according to popularized garden city principles, few complete new urbanist communities have yet been developed. Only a small proportion of projects called 'new urbanist' employ the full range of new urban principles or practices. Disseminating the principles in a fashion that produces complete exemplars of the potential of the movement is proving a challenge.

The Congress for the New Urbanism is justly proud of its accomplishments. Its newsletter regularly describes the growing numbers of members and projects. European organizations promoting new urban approaches have influenced government policy and programmes in the UK. Proponents of new urbanism see criticisms of the movement as unfounded and misguided, the product of ignorance or envy (see, e.g., Ellis 2002). They may justify the necessity of artifice in selecting building materials as a requirement to control production costs. They account for the lack of social mix as a product of the inability of the market to address social housing needs. They see the lack of affordability not as a result of new urbanism, but rather as demonstrating the need for even more new urban projects that will eventually bring down costs by enhancing supply. They criticize planning for forcing compromises and limiting the role of the designer to that of setting the rules for the built form. They make few apologies for failures to achieve the new urban dream.

After two decades of practice, new urbanism approaches definitely have their critics. Some dismiss the appeal of neo-traditional town planning, seeing it as nostalgic and artificial. They hold that plastic fences, non-functional front porches, fake balconies, and false dormer windows undermine the credibility of a movement that stresses the significance of authentic urban environments. They say that 'transit-ready' falls far short of 'transit-oriented'. Countless editorials have called the promoters of new urban approaches proselytizers, zealots, and missionaries for a cause. In many ways, we can see a war of values operating in the battle between 'conventional' (a.k.a. 'modern') approaches and 'traditional' (new urbanism) approaches.

Some commentators suggest that new urbanism has made less difference in practice than it might claim. Pyatok (2002:1) sees new urbanism as serving 'private developers who coopt their mission by simply repackaging suburban sprawl in more seductive "urbane" clothing, or public developers who too often trample on the lives of disadvantaged inner city communities'. In this sense, then, proponents may use new urbanist concepts and terms in professional discourse as a means to an end, as a way to gain approval for projects (Biddulph 2003b). New urban approaches simply serve, the argument goes, to make conventional development practices appear to have reformed (when they have not).

Social progressives find a dark side in new urbanism. 'Like urban renewal before it, these redevelopment processes are not serving to empower the residents of inner city neighborhoods, but are again displacing and scattering communities' (Milgrom 2002:9). Despite its progressive rhetoric, new urbanism facilitates gentrification, displacement, and segregation in some places. Its critique of the problems of the city has proven superficial, and its solutions do nothing to overcome the inherent inequity embodied in urban structure and replicated through new development.

While the critics may be growing in number, however, so are the converts. New urban approaches clearly appeal to urban élites. The new urbanists draw on historic precedents to generate community identity, provide local services, and mix people and uses. Governments have accepted the need for increasing urban intensity as a sustainable and smart strategy. New urban approaches are changing our cities and suburbs as the principles increasingly shape new construction projects. More attractive urban and suburban settings are resulting; more urban-oriented policy has been developed. We should not underestimate its impact on early twenty-first-century urban form even as we acknowledge its limitations.

CAN NEW URBANISM ACHIEVE ITS GOALS?

The proliferation of new urban projects has generated a cottage industry in testing new urbanist claims. While some research (e.g., Lund 2003) shows that particular urban forms may be associated with reduced car use or increased walking, empirical results to date prove conflicting. Of course, the new urban projects built are seldom old or large enough for effective testing in comparison with conventional developments. This has led many researchers to use older urban neighbourhoods (developed prior to the Second World War) as their 'new urban' proxy sample (e.g., Lund 2002): this method arguably proves flawed, since these established neighbourhoods are not strictly comparable to new developments. Their greater sense of community and propinquity for walking may correlate with other factors than design. It will be a few years yet before we have good data on whether the new urbanist projects achieve their objectives.

Some studies suggest that new urbanism has not yet succeeded in meeting some of its key goals. For instance, while some mixed-use projects have proven highly successful, making mix work in smaller developments seems a challenge (Grant 2002b). Getting people out of their cars and walking to work or shopping is not easy (Brown and Cropper 2001; Crane 1996; Handy 1996, 1992). When transit use has been declining for decades in some regions, finding strategies to promote transit-oriented development proves difficult (Cervero 1998). Efforts to manage growth through infill instead of greenfield development may force growth to leap-frog into other jurisdictions, or drive up the costs of development (Bae 2004). With mostly suburban projects built to date at fairly low densities, new urban approaches have not turned around the dominant trends in urban development. Thus, several of new urbanism's more ambitious objectives remain elusive at this stage.

We may argue, though, that new urbanist projects achieve their aims to create attractive and well-functioning spaces. New urban developments are generally beautiful, with a strong sense of place and character (see Figure 9.1). New urbanism adds value to land development, making it potentially lucrative for investors. Brindley (2003:61) notes that:

Place is something to which social identity can be attached and where a sense of security can be constructed. The aesthetics of place create particular, local meanings, so there is a demand for those qualities that make places distinctive and give them unique symbolic value.

New urbanism responds to that need for identity and place. In promoting traditional values, it acknowledges the value attributed to older settlements, and seeks to articulate the rules that may reproduce ancient meanings.

New urban approaches have clearly been successful in generating interesting examples of planned communities. The developments are photogenic and, like the factory towns and model garden cities before them, they generate considerable tourist traffic and journalistic interest. Many of the projects, especially by the better known firms, have a clear identity, character, and sense of place. New urbanism gives a locality a head start in generating sense of place by making it visually appealing and identifiable. People respond to those visual cues. While the intent may be linked to marketing directives, it also reflects an understanding of what contemporary cultures define as good places.

Efforts to generate a sense of place through mix, diversity, and self-containment have proven more of a challenge. Trying new development modes can be financially risky for developers, leading them to make choices that undermine the quality of places. Factors driving the location of commercial development limit the mix of uses in projects, making communities seem incomplete. Frantz and Collins (1999) emphasize how much Celebration residents wanted useful stores; instead they got

Figure 9.1 Lakelands street
Picture-perfect houses with picket fences line the streets in Lakelands, a new urbanist project in Gathersburg MD.

tourist-oriented retail that reinforced the sense of artifice about the place. In the Canadian new urban suburbs, commercial tenants have been slow to populate the infrastructure provided by developers anxious to create real places but consequently hobbled by the financial drain of carrying under-performing properties (see Figure 9.2).

Perhaps the greatest challenge for generating a sense of place through new urbanism comes from the realities of production development. Cities are built by countless companies competing with each other to produce structures for optimal returns. Some firms inevitably find a niche in the market mass-producing units at low cost, copying what works from each other, and following the rules set out by government and lenders. The resulting landscape is standardized and often mediocre. While leading edge developers may produce communities with a dynamic sense of place, most builders create spaces that look relatively similar from town to town. Even new urbanism projects suffer from this copycat syndrome.

One of the key premises of new urban approaches is that better urban form can contribute to the sense of community in an area. The findings on this are mixed. Brown and Cropper (2001) found a positive sense of community in the developments they surveyed. Brindley (2003) argues, though, that even traditional villages are becoming fragmented and incorporated into wider urban networks in the UK; hence, any hope of generating a local sense of community must recognize the larger social networks with which residents engage. Ultimately, sense of community develops over time through the interaction of community residents. Inasmuch as new

Figure 9.2 McKenzie Towne High Street
Designed to look as if they were built at different times, the shops of McKenzie Towne in Calgary AB reveal a post-modern interest in diversity. The second storey windows are false fronts, as the costs of putting apartments over the stores proved prohibitive. A block away, similar stores stood empty for more than 18 months awaiting tenants in a weak market.

urban forms can create spaces for people to interact socially, then they may support the growth of a sense of community. The same is true, however, for conventional developments, where community animation activities or shared recreational facilities may facilitate community-building (J. Martin 1996; Riger and Lavrakas 1981). If the good community is defined as a socially interactive setting, then new urbanist projects are arguably no better than other forms. For instance, squatter towns or trailer parks may reveal a strong sense of community among their members, not due to the form of the settlement but because of the character of the social networks generated within a community of interest. What new urbanism projects do provide, however, is a niche for segments of the middle and upper classes whose conception of community attaches to particular principles about urban form.

Seeing our own actions as having enormous potential to shape behaviour is an occupational hazard in the planning and design professions. We want to believe that we can change the world in ways we envision; we need to hope that our solutions can prove efficacious. The new urbanists certainly fall victim to such determinist thinking. They tend to define problems as technical, rather than social in nature, and affirm their ability to achieve social objectives through design strategies. New urbanism 'lives by an unswerving belief in the ability of the built environment to create a "sense of community"', says Talen (1999:1361). This environmental determinism implicit in new urbanism may be enough to warrant scepticism about its claims. While new urbanism certainly creates attractive places where people can meet, its bolder claims of generating community are not supported. If new urbanist projects do engender commonality amongst their residents, they do so primarily because of the homogeneity of those who buy into the projects. Residents self-select for their commitment to the principles that new urban projects reflect. Till (1993) argues that in marketing to regional concerns developers actually reinforce social divisions and increase homogeneity in neo-traditional projects. Like other upscale development movements, new urbanism has to worry about creating exclusive neighbourhoods that may facilitate social interaction amongst residents at the cost of excluding many categories of non-residents. Brindley (2003:62) suggests that in the urban village, 'The language of community building has become a moral code for policies intended to manage the resulting inequities and social conflicts' of the urban environment. As Frantz and Collins (1999:269) say: 'Building a community takes more than porches and potluck suppers. It takes time.' Even if a sense of community develops over the years, creating a well-rounded community of difference is not guaranteed.

MARKETING THE GOOD COMMUNITY

With the formation of the CNU, the new urbanism movement launched a full scale offensive to sell the idea in North America. Intense media coverage, workshops, videos, tool kits, and courses disseminate the message and the skills of building traditional communities. The movement's leaders – such as Andres Duany, Peter Calthorpe, James Howard Kunstler – give talks at professional conferences, write articles and books, and appear on television programmes. Several have taught courses on traditional urbanism at university architecture and planning schools: for instance, at Harvard, Miami, and Notre Dame (Ouellet 2004).

Professional planners' associations and developers' groups in America have become passionate advocates for smart growth. The American Planners Association (APA) and the Urban Land Institute have followed the CNU's lead in developing workshops, conference sessions, and tool kits, and producing books to promote the principles of new urban approaches. APA's new urbanism division hosted a session at the annual conference in April 2004 on 'Taking new urbanism to a new level'. Members of the division offered advice on how to promote new urbanism and smart growth (Chiarmonti 2004; Tirinnarri 2004). They described the kinds of images to show, words to use, and ideas to avoid. For instance, planners were told not to mention 'reducing car use' but instead to talk about 'choices in transportation'. The speakers advised planners to use 3-D imagery to illustrate new possibilities. To critique conventional forms, planners might show windowless school buildings, monumental traffic jams, and endless parking lots. To illustrate the benefits of new urbanism, planners should find images of nice streets with mixed housing, and street scenes of people in cafés and restaurants that are not too upscale. Packaging the message effectively has become an important strategy in the new urbanism movement, as it was in the garden city movement before it, and as is vital through the entire development industry.

DPZ, Calthorpe Associates, Krier Kohl, Urban Strategies, and other new urbanism firms rely on excellent graphic representation to entice clients. Their success in becoming lucrative design practices reflects their salesmanship as well as their considerable expertise and innovation. At times they have created 'histories' for the places they propose, to add flesh to the skeletons of the urban form. They have taken clients and planners on tours of the kinds of places they seek to emulate, to give a visceral feel for the anticipated benefits. For the most part, though, they use watercolours, paintings, 3-D visualizations, and drawings as sentimental marketing tools to persuade potential clients of the power of their concepts. Once the pretty pictures come out, resistance to new urbanism may melt away.

The development industry has taken note of new urban approaches, with mixed response. In the USA, the Urban Land Institute has been a strong advocate of new urbanism and smart growth, working hard to convince developers of the opportunities available. But the industry is quite pragmatic, and is also well aware of competing options like gated communities. Building groups in the UK have flagged their concerns about the viability of government policies that favour urban redevelopment over suburban expansion. Developers are always seeking an edge in a competitive market. A few have committed their practice to new urbanism, but many more cherry-pick the easy elements of it: attractive townhouses, front porches, and false dormers. Watered-down versions of urban villages and new urbanist neighbourhoods have become commonplace. Many contemporary suburbs sport hybrid notions, such as traditional porches and dormers on houses with front-attached garages in North American cities. As long as developers find a ready market for new urban communities and the aesthetic components that make new urbanism attractive, they will produce those elements. If the past is any indication, however, the development industry is as trendy as it is competitive. If the market for new urbanist product becomes saturated, the industry will seek new options.

THE IRONIES OF NEW URBANISM

As the garden city model was popularized through the twentieth century, its principles were gradually whittled down to a few standard development practices. The process of implementation gutted the theory of its important contributions. Practice is also taking a toll on the principles implicit in new urban approaches. Hence we can point to a series of ironies – discontinuities between what new urbanism promises in theory and what it can deliver in practice.

The new urbanists recognize that problems are complex, yet they often offer simple solutions. For instance, Krier (1984a:43) notes that the problems of the industrial city result in part from the concentration of power in the economy. His solution: smaller blocks! While the design options offered in new urban approaches may resolve some of the physical problems of the contemporary city, they fail to tackle the larger forces operating in the urban environment. Where the options may prove more complex, as in the regional city framework (Calthorpe and Fulton 2001), the solutions seem virtually impossible to implement in current political conditions. The new urbanists show little interest in the structures of power and inequality that undermine older districts or that disadvantage large numbers of urban residents. Instead, they look for simple, and sometimes wishful, strategies to guide development.

New urbanists advocate urban forms while building suburban enclaves. Are new urban approaches really anti-urban? Marcuse (2000) sees new urbanism as making it easier for the wealthy to flee to more attractive suburbs: he denies that it constitutes an urban approach. New urbanism may encourage fear of older urban areas with their chaotic medley of forms and people. Marcuse believes that, left to disintegrate, the ugly post-war suburbs might have driven people back to the city; instead pretty, insulated, middle class, new urbanist enclaves make it easy for people to choose the suburbs. At the APA conference, Ray Chiarmonti, planner for Tampa, discussed his commitment to new urbanism. He moved his family to Celebration because he wanted to live his values in a new urban community. Of course, then he had to commute 100 km every day by car to work in Tampa (Chiarmonti 2004; Franz and Collins 1999). Chiarmonti did not acknowledge any irony in his behaviour; instead, he talked about how easy he found walking to meetings at the Celebration school in the evenings. With few jobs and limited mix, new urbanism projects like Celebration are primarily dormitory suburbs. As Zimmerman (2001:260) says, the irony is that these developments that aspire to limit sprawl facilitate growth on the fringe.

New urbanists share the search for democratic communities and an egalitarian social vision while pandering to élite consumers. New urbanism asks for affordability and integration while building expensive enclaves. While we can cite some examples of projects with substantial components of affordable housing, most new urbanist developments are at the high end of local markets. New urbanists say that short supply forces up cost. But high costs also clearly reflect building quality and amenities in the communities. The first priority for the movement over the last 20 years has been to get projects built. Affordability, equity, and participation have been secondary. New urbanism has been slow to note the irreconcilable tension between democratic aspirations and market realities that may render some of their aspirations impossible.

New urban approaches call for an end to sterile and conformist suburbs while codifying new uniformity. The new urbanism critique of homogeneity and ugliness in contemporary suburbia is well-founded. Suburban landscapes are reproduced in similar forms across wide expanses. Each nation has its stereotypical modern standardized form. New urbanism seeks to change that by returning to traditional forms. While traditional styles also may have been quite standardized, indigenous building materials and designs responded to local cultural concerns. The problem in practice, however, is that new urbanism relies on codes to ensure the quality of new development. Rather than providing opportunities for innovation and experimentation, the codes entrench standards set by the designer when building projects. Furthermore, developers often employ designs and codes widely across landscapes, generating a new kind of uniformity. Designers call for flexibility while prescribing rigid rules. As Sitowski (2004:4) notes: 'The central irony here is that what once occurred "organically" must apparently now be codified.'

New urbanism invokes a concept of universal and timeless design principles while affirming the need for expert designers to set development parameters. If humanity shared consensus on a particular set of design principles that might hold across time and space, in a range of cultural contexts and conditions, then it seems reasonable to argue that we would not need expert designers to direct the design process. But new urbanism makes contradictory arguments here, assuring us that the role of the designer is more important than ever to identify and implement what we are simultaneously advised is a natural order.

New urbanists praise the public realm while creating private landscapes. The movement has made a significant contribution by drawing attention to the need to improve the public realm: streets, squares, and public buildings. At the same time, though, most of the projects created with new urban approaches are private spaces. In some, even the streets are private. 'Village centres' often comprise commercial spaces where unwanted individuals or groups can be excluded on management's call. The new urbanists criticize the planned unit developments of an earlier generation – places like Columbia, Maryland, or The Woodlands, Texas – even as they build a new generation of a similar product. They advocate for town centres while their new projects typically compete with and potentially undermine the economic viability of older commercial districts. The contemporary revival of downtown areas in many cities, with condominium apartment growth, owes much of its success to the principles of new urbanism. We might argue, though, that improving urban areas often involves turning them into exclusive entertainment districts for élites in our societies. The new zones become areas of tourism, or gentrified districts for affluent professionals. Public spaces increasingly feel like private enclaves, while private spaces are dressed up to appear as public squares.

Unless new urbanism is able to reconcile and transcend its ironies, its potential to become a more influential paradigm within planning theory may remain limited.

WHEN URBAN REALITIES INTERFERE

Were the problems of the contemporary city as simple as block size and parking location, then implementing new urban solutions would be commonplace. Unfortunately, they are not. We find a significant gap between the promise of new urban proposals and the realities of urban structure. Changing car-dominated development patterns that developed over the last century will not prove easy: vast areas of territory have infrastructure that suits scattered uses. Even strategies to retrofit suburban landscapes show little chance of short-term success. Many local governments have insufficient resources to build the infrastructure to support transit-oriented development (Filion 1996, 2004). Planners lack the tools necessary to implement many of the objectives of new urbanism and smart growth. Even the growing consensus on the appropriateness of the solutions is not generating rapid change.

Our economy has come to depend on growth trends intricately connected to automobility. Construction, road-building, just-in-time manufacturing, auto-servicing, insurance – all these industries depend on motorized vehicles and the land use patterns created to support them. Our dining pleasure depends on delicate fruits and fresh vegetables shipped north to our tables via these routes. The character of our cities reflects urban growth processes that transcend questions of form. In the American context, this has meant high building standards, exclusionary zoning, and racism, which Downs (1999) sees as all part of the same issue. In the Japanese context, it may mean high land costs, strong property rights, and a weak planning system (Sorensen 2002). Each country has a different constellation of factors that shape development patterns, and which often prove resistant to transformation through new urban approaches.

It seems safe to suggest that people like to live in the suburbs: that is where, of necessity, most people reside. And studies of suburban living reveal that most people are happy with their circumstances there (e.g., Baxandall and Ewen 2000; Talen 2001). Despite growing concerns about traffic, many people willingly commute long distances and for long periods every day. New urbanists remind us that most people have little choice: sprawl is the primary option being built. Others argue, though, that consumers are not as passive as that makes them seem. People do choose the places where they live. Our landscapes reflect cultural values and active choices made by households weighing the tradeoffs of varying options in the residential environment.

An American LIVES survey of 2000 home buyers indicated that the new urbanist package appeals to a small minority, perhaps 10 per cent, of consumers (T. Alexander 2000). Others say that perhaps one-third of American households might find new urbanism attractive (Audirac 1999). Most buyers in industrial societies, though, like larger lots rather than smaller. Given the level of affluence to afford lower density communities, they often choose less compact forms. In the North American context, few are willing to give up their cars, and their large front yards. In European and Asian cities, those who can are buying more cars and using them more frequently for commuting and shopping. The values these behaviours represent make selling new urbanism a challenge.

At the same time, though, new urbanism seems increasingly popular within a culture that values quick-paced, highly visual advertising, and short sound bites. Its visual quality makes great 'wallpaper' for reporters, and has helped it gain considerable exposure. The availability of the internet has also helped to spread the message and appeal of new urbanism. The quality of the places created in some new urban projects makes them popular leisure, touring, and shopping destinations.

People visit these communities and take home the message about potential future choices.

Many consumers appreciate the search for vibrant, diverse, attractive, and 'authentic' images and places. The lure of new urbanism reveals a reaction against the artificiality of contemporary culture. We see an authenticity in our urban past that eludes us in present-day communities. New urbanism picks up on the search as it works to create meaningful places. Hence Krieger (1991:38) can suggest in earnest that 'Windsor is designed to function as a real community', while we might ask how 320 rich families who love golf and visit a place for a few weeks a year make a 'real community'. By contrast, some new urbanists hold that the modernist public housing projects that accommodate the poorest families constitute dysfunctional places – not real communities – and hence warrant demolition. New urbanists define the messy landscapes of the inner city and the suburbs as artificial or dysfunctional while constructing myths about the authenticity of new traditional enclaves. They denigrate social networks formed outside their well-designed landscapes. In this sense, new urbanist understandings of community reveal élite values.

While we are unlikely to find complete consensus on what makes a good community, we see some common themes in popular culture. Many of these ideas resonate in an international context as well. People think of good communities as vibrant, diverse, dynamic, 'happening', attractive, and caring. Good communities are places where people feel comfortable and safe at all ages and stages of their lives. New urbanism ties into popular desires for real and good communities by offering design strategies that promise desired attributes and that complement governments' development objectives for urban improvement.

The development industry is continually engaged in creating neighbourhoods for a consumer society. Some elements in our communities have wealth to spend. New urbanism offers yet another option for conspicuous consumption: 'will that be a brownstone with an alley garage, a penthouse apartment overlooking the river, or a bungalow in a gated compound?' Like other upscale projects, new urban communities allow producers and consumers to code behaviour; ensure conformity, security and control; and protect property values and exclusive lifestyle choices. While the mid-twentieth century saw a period of growing working-class spending power, the late twentieth century generated double-income households, smaller families, and greater income disparity. Television had an enormous impact on leisure time, behaviour, and social expectations (Putnam 1995). Public policy has also shaped the ways in which we produce and consume landscapes; for example, in the 1940s through the 1960s government encouraged slum clearance; in the early twenty-first century, many governments seem ready to force front porches on housing consumers (Putnam 1995).

We might argue that new urbanist projects are not so different from gated enclaves. They represent variant forms by which the development industry seeks to cater to the needs and insecurities of urban élites through private communities. While new urban approaches focus on beauty and vibrancy, gated communities market security and privacy. The rapid proliferation of gated developments around the world indicates both the deterioration in urban living conditions and the significance of the appeal of separation in the consumer market. Rather than reforming societies that generate inequality and fear, these urban approaches facilitate spatial exclusion and socio-economic segregation. While new urbanism may frame itself as politically progressive, with its language of sustainable, healthy, smart, and environmentally responsible growth, it operates within a regressive system that reproduces inequality and alienation.

THE TRANSFORMATION OF PLANNING PRINCIPLES

As we examine the practice of new urbanism in various locations, we find networks of connections between scholars, practitioners, and patrons (Ouellet 2004). With the linkages enabled by contemporary communication and transportation technologies, design ideas can readily be globalized. Conferences, workshops, commissions, and competitions bring designers together to compare and share ideas. Publications, web sites, and email lists facilitate the dissemination of concepts and strategies across territories with radically different histories, and enable a degree of convergence of thinking about design principles unknown in an earlier era.

New urbanism began as an architectural paradigm and critique of planning. It developed within architecture as another periodic restoration of classical principles. In the late nineteenth century, Beaux Arts and city beautiful models developed in response to the problems of the ugly industrial city. In the twentieth century, new urbanism revived classical principles as a response to modernism. We might argue that disciplines since the Enlightenment naturally search for new approaches and theories to explain evidence and direct action. Theorists take advantage of constant opportunities to challenge contemporary dogmas. Some new theories and approaches catch on because of their clear applicability, and consequently set themselves up to be debunked as that practice is examined. This was the fate of modernism and the garden city. Now new urbanism faces the same test.

A relatively small group of self-selected visionaries, funded and supported by a few extraordinarily influential and powerful men, has managed to inspire a generation of practitioners and to shape development policy in many countries. The movement hardly had a democratic start, yet its ideas are spreading rapidly. Its leaders have become important cultural players. A small number of firms, like DPZ and Calthorpe

Associates, designs a large proportion of the projects in the USA (Al-Hindi 2001); in Europe as well, a few companies like Krier Kohl and Tagliaventi and Associates are very influential. The key architects in the firms may charge thousands for professional consultations and workshops. Despite the cost, cities line up to seek advice that they perceive may change their fortunes. The visionaries of new urbanism offer hope along with their technical advice. The lure of the new urban approaches reflects the desperate competition cities around the world face in their efforts to attract increasingly mobile global capital.

New urbanism constructs are increasingly embedded as *planning principles*, and are affecting practice in many if not most jurisdictions. This reality flags a fascinating question. Why have planners become so enamoured of a movement that has blamed them for the problems of contemporary cities? New urbanism criticizes planners for facilitating the industrial mode of urban production. Planners set the rules that guide the building of the city, employing a reductionist logic that fits an industrial model. Instead of operating from a normative vision, they served as policy writers and technicians. New urbanist critiques can be scathing in their attacks on planners. Why don't planners reject such disdain?

Planning has always involved a blend of the visionary and the technical. Until the 1960s, many planners were originally trained as architects, surveyors, or engineers. Most undoubtedly began their careers with a sense of vision – students usually do – only to find themselves in a practice that grew increasingly bureaucratic as it became more successfully entrenched within government. For most of the mid-twentieth century, planning was associated closely with modernism, forming part of the government apparatus that adapted cities to the needs of industry; the new urbanists describe it as pandering to the automobile. Planners gathered statistics, wrote plans, and implemented regulations. While a few planners drew images of ideal communities, most wrote policy, drafted maps, and prepared statistics to help decision makers run cities.

Coming from an architectural tradition, many new urbanists prefer visual presentation over the written word. Wyatt (2004) notes the disjuncture in the style of rhetoric and discourse in architecture and planning. The professions use different rhetorical strategies and engage a contrasting lexicon. Architectural training prepares its practitioners to convey ideas through images, and to generate novel solutions to contemporary problems. Planners learn to examine problems in political, social, and economic contexts, to use rational methods to work through options for solutions, and to write and interpret policy; they are cautious about imposing their own values on the decision process. It is not surprising that the professions would criticize each other, given their divergent world views; it *is* surprising, however, that the architectural critique had such an influence on a significant segment of planning practice.

Physical planning made its revival in late twentieth century at least in part because critiques within planning had never been suitably resolved. The failure of rational planning to account for the nature of practice or to provide appropriate normative guidance was well known by the 1970s (Friedmann and Hudson 1974; Hudson 1979). The search for alternative models to guide practice produced interesting theories, but not a unifying framework. Healthy communities and sustainable development theories certainly advocated a combination of approaches (design, policy, and regulations) that appealed to planners by fitting the principles that planners advocated while offering development options to address valid critiques of the profession. In the event, though, those theories became muddied and twisted by competing interests and hence could not make the necessary breakthrough to become a new legitimating paradigm for practice. Government retrenchment and cut-backs in the 1970s and 1980s created a context in which policy planning increasingly came under attack. Tired of facing blame for the problems of the cities, and anxious to find a paradigm that might reclaim some of the early vision and hope of the profession, many planners found the assertions of new urbanism attractive. In reaffirming the importance of physical planning, planners in some ways returned to the 'roots' of the profession – to the place where plans connect design and community – and put aside the difficult social and economic issues which they lacked the tools to resolve.

In the new urban image of the good community (spatial and social), the key principles of planning could be narrowed, renewed, and reinforced. A more clearly articulated idea of the public interest might be encapsulated within particular spatial solutions, with a presumed link to desired social outcomes. Ebenezer Howard's system of self-contained cities survived in the new paradigm, as did Raymond Unwin's commitment to craftsmanship and quality design. Camillo Sitte's principles of beautiful cities remains. Although turned inside out, Clarence Perry's neighbourhood unit provided the scale and combination of uses for the new urban paradigms. Jane Jacobs's notion of fine-grained mixing became a key premise. Christopher Alexander's timeless way of building offered important theoretical backing and implied universality. In recent years, the new urbanists have sought to integrate a few concepts from Ian McHarg (1969), and may talk about design with nature. Of course, new urban approaches have proven selective in their appropriation of concepts, terms, and theories from other models; connecting their claims to previous paradigms, naturally helped to strengthen their case.

New urbanism offered planners a neat package of planning principles. The proponents, with help from powerful patrons, built examples to show what new urban approaches could offer. Seaside became an early icon, while Celebration, Kentlands, and Poundbury provided test cases of larger-scale development. As Talen (2001) argues, new urbanism offered not only a changed building form but a new normative paradigm. Here was a different way of thinking about how cities might look and work, a return to a spatial and social image of the good community. As the movement gained media attention, it attracted planning converts in key places.

I have argued that new urban approaches clearly privilege the architect/designer/planner. The methods used in new urbanism seek to overcome the industrial mode of producing cities – through rules shaping development – to a system where the designer moulds the context through design codes. This privileging of design is not, however, always easy to sell to planners. The revival of physical planning, though, provided a mechanism for improving the prospects of design within planning itself. The 1990s saw more teaching of heritage conservation, physical planning, community design, and urban design in planning schools, as the market for new urbanism grew. By the mid to late 1990s, the planning schools produced design planners adept at articulating the role of the planner in new urban approaches.

Recently, new urbanism has developed ways to integrate its ideas and vision within the tools of planning. Early new urban projects relied on extensive private deed restrictions and covenants to protect the design vision of projects. The Seaside code and the Celebration pattern book offer well-known examples. In the early twenty-first century, both the CNU and the APA in the USA are heavily promoting the idea of form-based or context-based codes: a new generation of zoning. DPZ has pioneered the SmartCode, and seen it adopted in a few American jurisdictions. Municipal planners in some Canadian cities have transformed planning documents and design guidelines to promote new urbanism. Policy guidance notes in Britain encourage communities to adapt their plans and development controls. Such public codes allow planners to use mechanisms and processes they know to implement new ideas and principles. The new mechanisms may improve on private deed restrictions because as public regulations they apply to all property and can be changed with new needs and values. Planners play a more important role in the context of public codes than they could with private restrictions: they will write or at least implement the codes. Municipal coding can fully integrate new urbanism within planning as the new design paradigm.

When planning rules limited the mix or proximity of uses, planners could argue that we were ensuring health and safety in the city. In setting design codes, what over-riding public interest do we argue? The new planning principles go far beyond separating incompatible uses or ensuring traffic safety. Are we legislating taste in requiring classical lines, massing, and proportions? What do we see as the fundamental justification for our practice? In our search for a normative theory to legitimate our decisions and advice, we have found new urban approaches tempting, but as with our previous paradigms, those principles will prove subject to challenge and change through time.

THE ROLE OF THE PLANNER

Planners are highly motivated to contribute to positive urban futures. We want to end sprawl and create good places for people to live their lives. As a profession, we search for appropriate solutions to justify the authority we have to manage land use and form in cities. The last few decades created something of a fear of failure among planners, a desperate search for options that might control a market and culture that otherwise produces big-box retail, gated communities, garage-front streets, traffic grid-lock, and racial segregation. We want solutions that can work to create meaning and value in a context of globalization and mass production. We set a tall order for ourselves.

In wanting to justify our own role in government, we may implicitly promise more than we can deliver. We facilitate intensive visioning exercises in communities to generate ideas of what our cities might become, but then recognize our powerlessness to achieve the concepts. Can planners create the conditions for producing good communities? Despite the important role our profession plays in shaping development practices, in many ways planners' powers are limited. Planners cannot operate independently of the organizations, economies, and societies that employ us. In many ways, we are constrained to be followers rather than leaders. We may offer visions to our communities, but if the visions do not resonate with community values, we will not persuade local governments to adopt them.

The growth in new urban approaches reflects the need many planners feel for a sense of vision and mission within parameters that does not challenge the overarching framework within which we operate. Acknowledging the failure of rational planning left planners feeling insecure about the nature of their role in society. Dyckman (1983:8) argued that: 'If the field of community planning has set aside utopian ideas of the good society, it has paid for its profession with loss of substance and with the reduction of its theme of "reason" to instrumental rationality.' A think tank of Canadian planners pushed the profession to move beyond self-criticism to see the role of the planner as a visionary:

> it is essential that we take planning seriously – and be seen to take it seriously. There are real issues for planners to deal with – they cannot be taken lightly. A certain amount of scepticism may be useful in achieving perspective but should not replace our planners' sense of mission.
> (CIP Task Force 1982:11)

New urbanism filled the need planners felt for a paradigm that offered hope and prospects for positive futures rather than trenchant criticism that seemed to provide no viable strategies for action.

Although planning organizations hope that planners can be visionaries, for the most part planners' employment does not afford them such luxury. Planners more often play the role of enablers: finding ways to translate community values into policies, regulations, and codes to govern built form. In practice, public expectations of what is required to create good communities involve much more than issues of built form and land use. Planning cannot control many of the factors that generate problems in our cities. Thus results commonly fall far short of aspirations.

Governments are limited in their abilities to shape the future of our communities. Government can set directions that help to define the good community, but achieving targets proves a challenge in the context of adversarial democracy. Governments come and go as the public expresses disappointments with goals not achieved, or chases new hopes and promises. The 'good community' is a moving target, defined differently by competing interests and political parties. Achieving significant goals may take long term commitment that is not always available; instead, local governments face immediate concerns from short term issues. Moreover they have limited resources and imperfect knowledge of the future. They operate in a context where decisions made by large corporations often have more drastic impacts on cities than does government policy.

For new urbanists, government's role seems simple: plan flexibly, articulate vision, set and enforce design guidelines, facilitate participation, fund integrated social housing, create opportunities, make land available below cost, and reduce risks for developers. New urban approaches need governments as players. Without government support, new urbanism runs into road blocks that limit innovation and cause delays that may discourage developers. Achieving better futures, they argue, requires activist governments that will frame a context in which new urbanism can succeed. That means changing the rules of development. Moreover, government is expected to take responsibility for physical and social infrastructure investments, from transit systems to affordable housing. New urban approaches anticipate a more activist and interventionist government system; not surprisingly, they have generated a strong reaction from those associated with organizations that oppose large government and advocate free enterprise.

While a few local governments have become strong advocates of new urban approaches, often we see higher levels of government actively promoting new urbanism principles. State and national governments view strong cities as vital to national economic success. Economists of varying stripes increasingly describe urban regions as economic drivers, competing with global cities for expansion and growth. Governments want to harness urban regions to work towards achieving state ambitions. In some cases, this has been defined to mean smart growth or urban renaissance: economic development strategies with new urban approaches at their heart. As state enterprise, then, planning plays a key role in implementing these approaches to ensure urban competitiveness in the global market.

Some local governments have remained sceptical of new planning approaches. Local elected officials and staff have close dealings with a range of developers. They may not see the harm in conventional development forms that provide good sources of revenue and seem generally popular with home buyers and voters. Local governments may be willing to accept a range of options, including smart growth or urban villages concepts accompanied by incentives and funding programs (Biddulph 2003b). Short term success or failure may affect the commitment of local governments to new urban approaches. If projects work, others may follow, but if they fail, then conventional practices will likely resume, and planners may be forced to return to their roles as policy-writers and development technicians.

THE ROLE OF THEORY

Do new urban approaches contribute to or draw on planning theory? Do the principles of new urbanism derive from planning theory? Is new urbanism driving planning theory in new directions? Does new urbanism challenge key elements of contemporary planning theory or contribute to the development of a more resilient normative theory of planning?

I see few ways in which new urbanism connects to mainstream planning theory. New urbanism sidelines many of the key themes – power, communication, large-scale spatial patterns – that interest planning theorists. Many key planning theorists similarly say little about new urban approaches, focusing instead on questions of process and macro-level patterns. As studies of new urbanist practice become more common, we will likely see more discussion in planning theory about the ways in which new urban approaches have become so popular among practitioners, but in that context new urbanism would provide an object of investigation rather than a paradigm for promulgation.

New urbanism presents an analysis and options that ignore key elements of culture and political economy; in that sense, its theoretical assumptions may appear naïve to the planning theorist. Theorists are likely to find a theory that pins the problems of the city on planners and cars too simple. They ask: what about the economy of capital that drives urban development and creates both the conditions that generate sprawl and new urbanism and smart growth as potential solutions to it (Harvey 1997; Knox 1992, 1993)? What about changes in retail patterns, employment, and income that affect real estate preferences? What about changes in lifestyle, leisure, behaviour, household composition, and child-rearing that shape choices people make about where to live and how to use their automobiles? Understanding the city and the ways in which planning may affect its shape and function requires that we look at these larger questions to explain current circumstances and predict future possibilities.

As a practice, new urbanism has a particular appeal in the constellation of contemporary values and needs. This certainly makes its advocates anxious to afford it greater significance in the theory of the profession. To date, though, its proponents' eclectic efforts to develop theory have not proven effective. The search for timeless principles of design has left the new urbanists vulnerable to charges that they practise the kind of reductionist thinking they criticized in the modernists. The faith in universal design principles reassures designers: if there are timeless values, then designers who master them can make the right choices. They avoid risk that their theory or designs will be criticized in the future (as we currently challenge modernist thinking and forms). Yet even though classical principles of design continually (re)circulate in Western culture, we understand that they are culturally conditioned. They comprise one among a set of principles that gain popularity from time to time. Asserting through theory that particular values are universal and timeless has a normalizing, standardizing, and perhaps even a moralizing effect. Those who define the 'timeless principles' set the values. Their culture becomes mass culture, fixed in built form for generations. Such theory is arguably chauvinistic.

Is the new urbanism helping planners to define or redefine our understanding of the public interest and the nature of the good community? Talen (2001) says that achieving the common good requires new urbanist solutions. But the very notion that we might define the good community in precise or unitary terms must be challenged in an era of growing cultural diversity. Accommodating and supporting diversity represents an important theme in contemporary planning theory, especially in the work of Leonie Sandercock (1998). The concept of the public interest has become little more than an abstract rhetorical concept in new urbanism as it was in rational planning, useful as a rallying point but without clear substance. In the battle to make new urban approaches the norm, however, we find frequent appeals to a notion of consensus on best choices.

In some ways new urbanism presents an argument to contemporary planning theory about the values related to privilege and the nature of communication. New urbanism sees beauty, character, and connectedness as key to urban livability. By contrast, the most influential planning theorists focus on the values of democracy and equity. Contemporary communicative planning theory comes out strongly in favour of promoting the interests of the least powerful in society, a kind of updated radical planning approach. Authors like Healey (1997), Forester (1989), and J. Friedmann (1987) expose the inequalities and inefficiencies of practice in order to facilitate more participatory and collaborative planning processes. New urbanists do not often cite critical theory or communicative action in their writings. While they believe in the necessity of effective communication to achieve their goals, their understanding of the processes of communication differs from that of the theorists. New urbanists communicate to persuade, while planning theorists deconstruct communication to parse it for hidden messages and representations of power. New urbanists engage in collaboration to create local design solutions, whereas planning theorists look to collaborative methods to transform existing power dynamics and empower local people.

Planning theory and new urbanism treat space quite differently. New urbanists see space as a manipulable receptacle for creating venues for social activities and a product for generating property values. They do not problematize space in the way that theorists do. Planning theorists are interested in the ways in which spaces interact with and reflect developments within social and economic processes. For the new urbanists, space has a static quality: if its features are good, then it should be maintained forever; if its features are bad, then it warrants replacement. For planning theorists, space has a dynamic nature: constantly in flux as it represents the interactions of forces that planning seeks to manage. Planning theorists reject the simple representations of space that suit the new urbanists.

New urbanists wonder why planning theory seems reluctant to embrace new urban approaches as providing an underlying set of principles for practice. New urbanism practitioners certainly talk about many issues of concern to theory. Why has new urbanism not been better integrated into the theory taught in planning schools and written in standard planning theory texts? From the perspective of planning theory, new urbanism simplifies or ignores difficult questions of practice. In working to facilitate urban development, new urbanism appears to serve the interests of power. Its practitioners communicate through an élitist discourse that disempowers and coopts community members. Its promotion of universal principles facilitates globalization and standardization. Its focus on local options ignores larger spatial issues and questions of political economy. Its view of urban problems is overly circumscribed.

Planning theorists wonder why practitioners have little interest in theory: this question has animated the theory stream at meetings of the Association of Collegiate Schools of Planning and the Association of European Schools of Planning. Why do planners seem uninterested in reading theory as a guide to practice? How can theorists make theory relevant and engaging for practitioners? Instead of streamlining issues for planners, theory has the bad habit of often making problems seem so complicated and intractable that it may undermine practitioners' hopes of generating change. Theory has the ability to paralyse, either because it is written in a jargon impenetrable to all but the keenest doctoral students, or because it effectively ignores the question of immediate application. Practitioners seek options that they can implement over the short term in the hope of long term benefits. While they may find an analysis of the discourse of planning amusing, they do not know how to use such tools to change their community in ways that will make things better. By contrast, new urbanism offers a set of simple rules that practitioners can implement immediately. Its theory is minimal; its allure is compelling.

As planners do we have the tools needed to deal with issues of poverty, inequality, and marginalization? We find that we cannot easily achieve the goals of diversity, affordability, and social integration, whatever our theory or practice. Financiers and neighbouring communities often resist our strategies of mixed use and transit-oriented design, despite widespread professional acceptance of them in both theory and practice. Our record at responding effectively to the social issues of the day seemed good in the early years of planning, when we focused on promoting public health and separating conflicting land uses. How well can we respond today to the key issues of our times? Do we share sufficient consensus about the problems of the city and how planning can help address them? While some see sprawl as the critical concern, others argue that inequality is the greatest challenge. Whichever of these issues we focus on, however, planners need theory and tools that give us some level of assurance that we know how to respond effectively.

PLANNING AND THE SEARCH FOR THE GOOD COMMUNITY

The history of urbanism shows an enormous range of design options and preferences through time. While some people favour urban forms that enclose and define spaces, other peoples at other times resist enclosure. We can find low density cities, like those of the ancient Maya (Hardoy 1968), and very high density cities, like those of contemporary Asia. Tastes in urban design change over time. Needs in the urban environment change. Design has to be able to adapt.

New urban approaches have become popular because they seem to offer practical solutions to contemporary problems. Beauregard (2002:182) says that 'New Urbanism is less a journey to a safe and stable new world than a complex negotiation of a deeply divided present'. New urban approaches appeal to our sense of what a good community might look like and how it might function. They produce attractive places with a strong sense of place and generally good market potential. New urbanism manages to be at the same time nostalgic about a civilized past and hopeful about the future. It draws on useful lessons from past glories to derive principles to give cities images of a better future. It promises transformation.

Can new urbanism produce good communities? The advocates of new urbanism certainly believe that it can. Appropriate policies and regulations can create places that have attractive and functional physical form: suitable containers for the good community. For many, those physical elements are more than enough for us to expect the state to produce: beyond that, society must fill up spaces with the other attributes necessary for good communities. Practice to date offers less evidence that new urbanism can produce the social or economic dimensions of good communities. If we believe (and not everyone does) that good communities should be inclusive, empowering, democratic, affordable, adaptable, and environmentally responsible, then attractive physical places will not be enough. As every planning approach of the last century has discovered, no simple formula guarantees equity. Delivering beauty will always prove easier than ensuring that everyone can find a place.

As planners, have we developed consensus on what planning the good community might entail? New urbanism has its adherents, but despite its growing importance, not all of its premises are universally accepted. Government policy statements widely accept principles of mixed use and compact form, for instance, but major differences of opinion remain over key issues such as the place of nature in the city.

New urban approaches have tended to use villages or small towns as models for urban planning at the same time as they promote urbanism – something normally associated with larger cities. Krier has tried to resolve this tension by arguing for urban quarters, while Duany and his colleagues focus on the neighbourhood as the basic urban unit. New urbanism thus sees the good city as principally an accumulation of attractive building blocks united by common physical attributes, and effective open space and transportation networks.

The good community, of course, must be more than the sum of its parts. Its neighbourhoods derive their healthiness and sustainability in part from the larger urban fabric and the connections between them. Optimistic neighbourhoods are knit within an integrated and well-functioning urban network. If authorities accept the need for gated enclosures, as so many local and state governments have done, then they admit that they have not created an urban network within which people can find comfortable communities. Good communities constitute healthy, safe, and efficient places, with caring people. Only a cultural context that provides stability and security for the population at large can create them.

How important is the role of urban form in contributing to the good community? That proves difficult to measure. People like attractive places, but they define a wide array of places as attractive. We find little consensus on the shape of the good community over time or space. While classical principles certainly have their adherents, they are not universally loved. The good community can come in a variety of shapes. What might be common about the concept of the good community is the state of mind and body of its inhabitants, rather than the shape of its streets and squares. That is, in the good community, people can be healthy, happy, and productive.

Future prospects

> In every age someone, looking at Fedora as it was, imagined a way of making it the ideal city, but while he constructed his miniature model, Fedora was already no longer the same as before, and what had been until yesterday a possible future became only a toy in a glass globe.
>
> (Calvino 1974:32)

New urbanism approaches are influencing contemporary practice in many countries, but can we expect them to remain favoured solutions for producing good communities? It seems unlikely. New urbanism first became popular in the USA and Canada because it responded to particular North American concerns: sprawl, heritage destruction, and automobile culture. The compact city and urban villages appealed to the perceived need to contain new development in a way that reflected urban traditions in the UK and Europe. In Asia, new urban approaches encounter urban residents anxious to expand the amount of domestic space they have, to have more say about urban development, to improve air and water quality, and to increase automobile ownership and use. In many parts of the world today the reality of insecurity leads to widespread gating while high land values drive high-rise housing projects. In many Third World cities, compact form and mixed use are already commonplace, but people need access to basic infrastructure and secure tenure. In other words, each region, and perhaps each city, has its own history and problems and must find its own strategy for facing its challenges. Some approaches may lead to good communities; others will not. Only history will define which is which. One thing seems clear: new urbanism will not work everywhere.

New urbanism is a reality of urban development in many Western societies in the early twenty-first century: how durable will its appeal prove? If asked to foretell the future, I would predict that new urbanism may be relatively short-lived. Classical principles had a major impact on Greek and Roman city building, and on church architecture in Europe: in these cases, the longevity of the form reflected the nature of social control and authority in the societies involved. In an open society where competition of ideas adds value to particular forms, design movements have short half-lives. The Beaux Arts revival and city beautiful movement lasted only a couple of decades, influencing public and corporate architecture, but not penetrating far into popular culture. Other twentieth-century urban movements, like the garden city and modernism, flourished for several decades before being stripped of their principles and reduced to low-cost mass-produced urban forms. Urban traditions may last where innovation is not valued and rewarded, but in the contemporary urban environment that is rarely the case. As the problems that capture public attention shift, so do ideas about suitable strategies for dealing with them. We already see evidence of new urban approaches reduced to a few marketable elements. Moreover, security communities are increasing more quickly than new urban projects. A larger proportion of new developments feature gates than those that represent full-blown new urbanist principles. Rather than presaging a return to urbane landscapes, new urbanism may represent a vain effort to stave off the urban fragmentation that prevails in the early twenty-first century.

Despite efforts to define new urban communities as healthy, sustainable, caring, livable, and moral, practice reveals the shortcomings of new urban approaches. The transformation in lifestyle required by new urbanism demands value change. The garden city model thrived through the twentieth century because it supported and reinforced mainstream cultural values, and found effective ways to articulate those values in built form. New urbanism, though, takes élitist values and seeks to disseminate them. In that it runs into the same problem that the environmentalists and Green Party advocates face: that is, mainstream values prove resistant to conversion. Values can change over time, but they do not do so just because designers offer alternative development models or political parties argue for responsible decision-making. The key values driving urban form in the contemporary period include capital (property values and return on investment), security (fear of difference and crime), and identity (need for status and self-actualization). Urban form movements, like gated communities, that respond to those values are the ones thriving in the period.

New urbanism has sought to articulate a new vision and mission for planning. In that it has made a significant contribution to the profession, and offered hope to practitioners eager to create more vibrant and attractive cities. It provides the ideological basis for urban revitalization and redevelopment activities through the Western world, and is influencing developments elsewhere. As a set of theoretical constructs, new urbanism may have its limitations, yet it has helped to unify urban practitioners. That means, of course, that practice not only implements new urbanism principles widely but also exposes its limitations and challenges its proponents to enhance their theory.

We can expect planners to continue to look for the one big theory that can explain all, predict all, and offer guidance for practice to create good communities. We can also safely predict that we are not likely to find such a model. The nature of the planning profession is to help communities respond to changing economic, technological, and cultural trends. In dynamic societies, that requires constant adaptation to local circumstances and international conditions. The good community remains an important but elusive target.

NOTES

CHAPTER 1

1 Ward (1998) suggests that cities around the world market themselves as vibrant and entertaining. Mixed use, compact form, and urban revitalization are common strategies for encouraging investment and immigration.

2 Of course, the idea of the nuclear family with a stay-at-home mother as the basis for the garden city model was a myth. Working class women have always worked outside or inside the home, whether as domestic, agricultural, or industrial labour. Furthermore, arguments that suggest that extended families were the most common type of household before the early twentieth century are not supported in history: the nuclear family appeared in European society as the dominant household type by the late middle ages (Huppert 1986).

3 Euclidean geometry postulates that only one line may be drawn through a given point parallel to a given line. Zoning draws lines around areas of particular uses. It does not employ Euclidean geometry in any way other than for rhetorical purposes.

4 Their rapport may have encouraged Duany to add an accent to his first name in recent years. Although I write 'Leon' throughout for Krier's first name, he actually spells it 'Léon'. In recent years, Duany gives his first name as 'Andrés'.

5 Al-Hindi and Staddon (1997:350) suggest that Seaside constituted a 'cultural revolution against urban modernism itself'. That may overstate the case, but it certainly revolutionized American thinking about settlement form.

6 Upton (2002:254) describes his use of the term *praxis* 'in the simple sense of an alternative source of knowledge to *episteme*'. That may not be perfectly clear to the average reader. I believe praxis implies knowledge gained from doing. Flyvberg (2001) employs the even more esoteric notion of *phronesis* to suggest practical wisdom. By employing Greek terms for simple concepts, theorists give theory a bad rap and invoke unnecessary headaches.

7 Of course, the Japanese love Tokyo, in spite of its chaos (or maybe because of it). Young people aspire to study and live in Tokyo: a job in Tokyo is a dream for many. While grand boulevards may account for the appeal of Paris to the French, social and economic energy and activity make Tokyo exciting. Those who love the classical lines of Paris have trouble understanding the appeal of Tokyo, but the Japanese do not.

CHAPTER 3

1. As Cliff Hague (personal communication) notes, Krier overlooks the reality that the city relies on the suburbs to house the labour force that the city requires to survive. If suburban dwellers are 'parasites' we might hold fiscal decisions rather than urban form responsible for the situation.
2. Urban villages are similar in scale to Perry's neighbourhood units. Although urban villages are smaller than garden cities, they seem to aim for the same blend of town and country features in a relatively self-sufficient community that motivated Howard.
3. Few, if any, of the major players in the new urbanism movement attended or graduated from planning schools. This may explain why they seldom cite planning theorists.
4. Zimmerman (2001) describes Prairie Crossing, near Chicago, as a suburban fringe new urbanist development that claims to conserve traditional ecology on the site. As he argues, though, the project reveals the developers' intent to use ecology and nature as marketing mechanisms.

CHAPTER 4

1. Weston financed the early Congress for the New Urbanism meetings (Thompson-Fawcett 2003b) and thus played a major role in facilitating the development of the movement.
2. Officially the project is called 'The Kentlands', but I use the simpler form of 'Kentlands' throughout.
3. Cliff Hague (personal communication) notes that UK planners try to restrict service uses in High Street premises in order to keep the streets animated.
4. Although I hesitate to jump to conclusions – since some infants could have been adopted – I could not help inferring that at least some of the women probably worked as nannies or housekeepers to white families in the neighbourhoods.
5. Housing and Urban Development Secretary Henry Cisneros was one of the original signatories to the CNU charter in 1996 (Bohl 2000).

CHAPTER 5

1. I'm grateful to Cliff Hague for clarifying the organization of planning guidance in the UK for me.
2. Steuteville (2003b) disputes the logic of arguments that new urbanism encourages crime and provides anecdotal evidence that crime is low in projects such as Celebration.

CHAPTER 6

1. Many Japanese completely and resolutely avoid association with any of the values or practices of the discredited régimes that led the country into the Second World War. Hence, reassertion of traditional architectural values and styles may fall within

the realm of dangerous and reactionary thinking. In East Asia generally, perhaps modernist approaches distance contemporary practice from militaristic and imperialistic traditions.

CHAPTER 7

1   In the mid to late 1990s, provincial governments in Ontario, Quebec, and Nova Scotia decided to amalgamate their largest cities with the towns and suburbs surrounding them. The same process had happened a generation earlier for cities like Winnipeg and Calgary. Regional governments were already in place in many other municipalities since the 1960s to coordinate planning and economic development activities. Provinces created regional cities to facilitate competition in a global economic market.
2   For example, the province of Nova Scotia borrowed federal money to buy land identified in a regional study (Coblentz 1963) as appropriate for future housing development. The provincial housing department then serviced land in those areas and sold lots at cost. This programme, which continued until the 1990s, played a major role in moderating the costs of starter housing in the province.
3   Parts of Ontario have two-tier local governments, with cities, towns, and villages within regions. Both levels have some planning authority. Normally local plans must conform to regional plans.

CHAPTER 8

1   In some ways, design codes emulate the sumptuary codes of an earlier era. Sumptuary codes, as Douglas and Isherwood (1996) explain, allow goods to function as part of an information system identifying status. In the middle ages, various classes were identified by their clothing; today, housing and automobiles have become more significant class markers.
2   Machiavelli's (1950 [1513]) enlightened despot certainly comes to mind.
3   We already saw classical and traditional principles fall out of favour after the Second World War, in large part due to a backlash against the fascist and militaristic régimes that had promoted them in Germany, Italy, and Japan. Design régimes that become linked to particular value or political positions are vulnerable when times change.
4   Cliff Hague (personal communication) notes that Margaret Thatcher's government in the UK issued a circular in 1980 that advised local planners that design matters were to be left to developers and their customers. Applications could not be rejected on design grounds.
5   Mendez (2005) argues that new urbanism could build on the cultural preferences of Latinos for compact form, but Day (2003) indicates that it has yet to do so effectively.

BIBLIOGRAPHY

30-A.com. (2004) 'Explore and discover WaterColor'. Online. Available at http://www.30a.com/discover/watercolor.asp (Accessed 7 October 2004).

Adler, J. (1994) 'The new burb is a village', *Newsweek* (26 December) 124(26): 109.

Ajax, Town of (1999) *A6 Community: Urban Design Guidelines*, Prepared by Viljoen Architect, Green Scheels Pidgeon Planning Consultants, Ajax, Ontario.

Alden, J. and H. Abe (1994) 'Some strengths and weaknesses of Japanese urban planning', in P. Shapira, I. Masser and D. Edgington (eds), *Planning for Cities and Regions in Japan*, Liverpool: University of Liverpool Press, pp. 12–24.

Aldrick, P. (2004) 'Wimpey frustration builds', *The Daily Telegraph* (London) 29 July, p. 32.

Alexander, C. (1979) *The Timeless Way of Building*, New York: Oxford University Press.

Alexander, C., S. Ishikawa, M. Silverstein, with M. Jacobson, I. Fiksdahl-King and S. Angel (1977) *A Pattern Language: Towns, buildings, construction*, New York: Oxford University Press.

Alexander, E. (2003) 'Havana charrette – human rights', PLANET, Online posting. Available email: PLANET@LISTSERV.BUFFALO.EDU 13 June 2003.

Alexander, T. (2000) 'Suburbia observed', *Urban Land* 59(7): 40–7.

Al-Hindi, K. Falconer (2001) 'The new urbanism: where and for whom? Investigation of a new paradigm', *Urban Geography* 22(3): 202–19.

Al-Hindi, K. Falconer and K. Till (2001) '(Re)placing the new urbanism debates: towards an interdisciplinary research agenda', *Urban Geography* 22(3): 189–201.

Al-Hindi, K. Falconer and C. Staddon (1997) 'The hidden histories and geographies of neotraditional town planning: the case of Seaside, Florida', *Environment and Planning D: Society and Space* 15: 349–72.

American Planning Association (APA) (2004) 'Codifying new urbanism'. APAStore. Online. Available at http://www.planning.org/bookservice/description.htm?BCODE=P526 (Accessed 14 September 2004).

Anderson, K. (1991) 'Old-fangled new towns', *Time* 137(20): 52–5.

Angotti, T. (2002) 'NU: the same old anti-urbanism'. *Planners Network* 151: 18–21.

Archer, J. (1988) 'Ideology and aspiration: individualism, the middle class, and the genesis of the Anglo-American suburb', *Journal of Urban History* 24(2): 214–53.

ArchNewsNow (2002) 'Exhibition/Award: Borneo Sporenburg residential waterfront, Amsterdam. November 25', ArchNewsNow.com Online. Available at http://www.archnewsnow.com/features/Feature90.htm (Accessed 17 October 2004).

Arendt, R. (2001) 'Sprawl and obesity', *Planning* 67(7): 34–6.

Atkins, R. (2003) 'The gate escape', *The Christian Science Monitor*, 21 September, B1.

Audirac, I. (1999) 'Stated preference for pedestrian proximity: an assessment of new urbanist sense of community', *Journal of Planning Education and Research* 19: 53–66.

Audirac, I. and A. H. Shermyen (1994) 'An evaluation of neotraditional design's social prescription: postmodern placebo or remedy for suburban malaise?', *Journal of Planning Education and Research* 13: 161–73.

Audirac, I., A. Shermyen and M. T. Smith. (1990) 'Ideal urban form and visions of the good life: Florida's growth management dilemma', *Journal of the American Planning Association* 56(4): 470–82.

Auffhammer, M. (2004) 'China, cars and carbon', Giannini Foundation of Agricultural Economics. Online. Available at http://www.agecon.ucdavis.edu/outreach/areupdatepdfs/UpdateV7N3/V7N3_4.pdf (Accessed 20 August 2004).

Aurbach, L. Jr (2000) Online posting. Available email PRO-URB@listserv.uga.edu (20 October 2000).

A Vision of Europe (AVOE) (2004), 'New civic architecture: the ecological alternative to suburbanization', Triennale IV of Architecture and Urbanism, Bologna 2004. Online. Available at http://www.avoe.org/TRIENNALE4index.html (Accessed 10 April 2004).

Bae, C.-H. C. (2004) 'Cross-border impacts of a growth management regime: Portland, Oregon, and Clark County, Washington', in A. Sorensen, P. Marcotullio and J Grant (eds), *Toward Sustainable Cities: East Asian, North American and European perspectives on managing urban change*, Aldershot: Ashgate, pp. 95–111.

Baetz, R. (1997) 'Back to the future of neighbourhoods', *The Hamilton Spectator*, 16 September, A11.

Banai, R. (1998) 'The new urbanism: an assessment of the core commercial areas', *Environment and Planning B: Planning and Design* 25: 169–85.

—— (1996) 'A theoretical assessment of the 'neotraditional' settlement form by dimensions of performance', *Environment and Planning B: Planning and design* 23: 177–90.

Barber, J. (2004) 'Instant modernity', *The Globe and Mail*, 23 October, R1–3.

—— (1997) 'Where the American dream lives on', *The Globe and Mail*, 4 June, A2.

—— (1995) 'Captain Cook couldn't find his way to my door', *The Globe and Mail* (Toronto Metro Edition), 15 February, A6.

Barnett, J. (1986) *The Elusive City: Five centuries of design, ambition, and miscalculation*, New York: Harper and Row.

Barstow, D. (2001) 'Envisioning the future in a fortress New York', *The New York Times*, 16 September, Online. Available at www.nytimes.com (Accessed 17 September 2001).

Baum, H. (1999) 'Forgetting to plan', *Journal of Planning Education and Research* 19: 2–14.

Baxandall, R. and E. Ewen (2000) *Picture Windows: How the suburbs happened*, New York: Basic Books.

BBC News (2003) 'UK commute "longest in Europe"', *BBC News World Edition*, Online. Available at http:// news.bbc.co.uk/2/hi/uk_news/3085647.stm (Accessed 5 August 2004).

Beasley, L. (2004) 'Working with the modernist legacy: new urbanism Vancouver style', Address to the Congress for the New Urbanism, Chicago, June. Online. Available at http://www.cnu.org/pdf/2004presentations/beasley.doc (Accessed 28 December 2004).

Beauregard, R. A. (2002) 'New urbanism: ambiguous certainties', *Journal of Architectural and Planning Research* 19(3): 181–94.

Benevolo, L. (1967) *The Origins of Town Planning*, Cambridge, MA: MIT Press.

Bentley Mays, J. (2001) 'Homes sweet homes', *Toronto Life* 35(8): 114–20.

—— (1997) 'The high priest of new urbanism', *The Globe and Mail* (Toronto), 8 March, C17.

Bergen, B. (1998) 'Open house will cull ideas for CFB Calgary', *The Calgary Herald*, 19 June, B10.

Berger, P. and T. Luckman (1966) *The Social Construction of Reality: A treatise in the sociology of knowledge*, Garden City, NY: Doubleday.

Berridge Lewinberg Greenberg and Associates (1996) *The Integrated Community: A study of alternative land development standards*, Ottawa: Canada Mortgage and Housing Corporation.

Berton, B. (2004) 'Cultured Pearl', *Urban Land* 63(6): 47–52.

Biddulph, M. (2003a) 'Implementing the urban village concept: lessons from practice', *Urban Design International* 8: 1–3.

—— (2003b) 'The limitations of the urban design village concept in neighbourhood renewal: a Merseyside case study', *Urban Design International* 8: 5–19.

Biddulph, M., B. Franklin and M. Tait (2003) 'From concept to completion: a critical analysis of the urban village', *Town Planning Review* 74(2): 165–93.

Birch, E. L. (1980) 'Radburn and the American planning movement: persistence of an idea', in D. Krueckeberg (ed.), *Introduction to Planning History in the United States*, New Brunswick, NJ: Center for Urban Policy Research, pp. 122–51.

Birmingham Grid for Learning (2003) 'Birmingham's Children and Young People's Strategic Plan (2004)-2010', *Birmingham's Children and Young People's Strategic Partnership*. Online. Available at http://www.bgfl.org/uploaded_documents/Childrens_Strategic_Plan_28_Sep_03.doc (Accessed 2 November 2004).

Blakely, E. J. and M. G. Snyder (1997) *Fortress America: Gated communities in the*

United States, Washington DC: Brookings Institution and the Lincoln Institute of Land Policy.

Blair, T. (2001) 'Improving your local environment'. Prime Minister's Speeches. 24 April 2001 Online. Available at http://www.number-to.gov.uk/output/Page1588.asp (Accessed 5 September 2004).

Blumer, H. (1969) *Symbolic Interactionism: Perspective and method*, Berkeley: University of California Press.

Bodenschatz, H. (2003) 'New Urbanism and the European perspective – presumption, rivalry or challenge?' in R. Krier, *Town Spaces, Contemporary Interpretations in Traditional Urbanism*; Krier – Kohl – Architects, Introduction by M. Graves, Birkhäuser Verlag, Basel, pp. 266–79. Online. Available at http://www.ceunet.de/newurbanism.htm (Accessed 11 August 2004).

—— 'New urbanism in the USA and urban design reform in Europe: common ground, differences and contradictions', Paper given at the World Congress of Architecture, Berlin, 22–26 July. *Resource Architecture* (conference proceedings) 2002: pp. 100–102.

Bohl, C. (2002) *Place Making: Developing town centers, main streets, and urban villages*, Washington: Urban Land Institute.

—— (2000) 'New urbanism and the city: potential applications and implications for distressed inner-city neighborhoods', *Housing Policy Debate* 11(4): 761–801.

Bonavia, D. (1978) *Peking*. The Great City Series, Amsterdam: Time-Life Books.

Bookout, L. (1992a) 'Neotraditional town planning: A new vision for the suburbs?', *Urban Land* 51(1): 20–6.

—— (1992b) 'Neotraditional town planning: Cars, pedestrians and transit', *Urban Land* 51(2): 10–15.

Bourne, L. S. (2001) 'The urban sprawl debate: myths, realities and hidden agendas', *Plan Canada* 41(4): 26–8.

Boyer, M. C. (1983) *Dreaming the Rational City: The myth of American city planning*, Cambridge, MA: MIT Press.

Breheny, M. (1992) 'The contradictions of the compact city: a review', in M. Breheny (ed.), *Sustainable Development and Urban Form*, London: Pion, pp. 138–59.

Bressi, T. (1994) 'Planning the American dream', in P. Katz (ed.), *The New Urbanism: Toward an architecture of community*, New York: McGraw Hill, pp. xxv-xlii.

Brindley, T. (2003) 'The social dimension of the urban village: a comparison of models for sustainable urban development', *Urban Design International* 8: 53–65.

Brophy, P. C. and R. N. Smith (1997) 'Mixed-income housing: factors for success', *CityScape: a Journal of Policy Development and Research* 3(2): 3–31.

Brown, B. and V. Cropper (2001) 'New urban and standard suburban subdivisions: evaluating psychological and social goals', *Journal of the American Planning Association* 67(4): 402–19.

Buder, S. (1990) *Visionaries and Planners: The garden city movement and the modern community*, New York: Oxford University Press.

—— (1967) *Pullman: An experiment in industrial order and community planning 1880–1930*, New York: Oxford University Press.

Burchell, R. (2004) Presentation in workshop, Smart Growth and Alternative Development Studies, American Planning Association Conference, Washington DC, 26 April 2004.

Calgary, City of (2004a) 'McKenzie Towne residential community market profile', Online. Available at http://content.calgary.ca/CCA/City+Hall/Business+Units/ Assessment/ McKenzie+Towne.htm (Accessed 14 October 2004).

—— (2004b) 'McKenzie Lake residential community market profile', Online. Available at http://content.calgary.ca/CCA/City+Hall/Business+Units/Assessment/McKenzie+ Lake.htm (Accessed 14 October 2004).

—— (2000) 'McKenzie Towne community profile', 'McKenzie Lake community profile', Calgary: Land Use and Mobility, Planning Policy Department (pamphlets).

—— (1995a) *Transportation Plan*, Calgary: Transportation and Planning and Building Departments.

—— (1995b) *Sustainable Suburbs Study*, Calgary: Planning and Building Department.

Calthorpe, P. (1994) 'The region', in P. Katz (ed.), *The New Urbanism: Toward an architecture of community*, New York: McGraw-Hill, pp. xi–xvi.

—— (1993) *The Next American Metropolis*, New York: Princeton Architectural Press.

Calthorpe, P. and W. Fulton (2001) *The Regional City: Planning for the end of sprawl*, Washington: Island Press.

Calvino, I. (1974) *Invisible Cities* (trans. W. Weaver), New York: Harcourt Brace Jovanovich.

Campbell, D. (2004) 'RIBA goes to town on radical revival', *The Guardian*, 31 July. Guardian Unlimited. Online. Available at http://society.guardian.co.uk/urbandesign/ story/0,11200,1273271,00.html (Accessed 7 August 2004).

Canada, Government of (1990) *The Green Plan*, Ottawa: Environment Canada.

Canada Mortgage and Housing Corporation (CMHC) (2004) 'The "Kings Regeneration" initiative', *Residential Intensification Case Studies*, Online. Available at http:// www.cmhc-schl.gc.ca/en/imquaf/hehosu/sucopl/loader.cfm?url=/commonspot/ security/getfile.cfm&PageID=64617 (Accessed 12 October 2004).

Canadian Institute of Planners (CIP) Task Force (1982) *CIP Task Force Report*, Ottawa: Canadian Institute of Planners.

Cervero, R. (1998) *The Transit Metropolis: A global inquiry*, Washington DC: Island Press.

—— (1986) *Suburban Gridlock*, New Brunswick, New Jersey: Center for Urban Policy Research, Rutgers University.

Chadwick, J. (1976) *The Mycenaean World*, Cambridge: Cambridge University Press.

Cherry, G. (1996) 'Bournville, England', *Journal of Urban History* 22(4): 493–508.

Chiarmonti, R. (2004) 'Taking new urbanism to a new level'. Paper delivered at the American Planning Association conference, Washington DC, April, 2004.

Chidley, J. (1997) 'The new burbs: after 50 years Canadians are re-thinking the suburban dream', *Maclean's* 110(29): 16–21.

Chiotti, P. (1992) 'Forbidden city: the once folksy town of Hidden Hills has become an affluent gated enclave that doesn't always take kindly to outsiders', *The Los Angeles Times*, 9 February, B3.

Chiras, D. and D. Wann (2003) *Superbia! 31 ways to create sustainable neighborhoods*, Gabriola Island BC: New Society Publishers.

CNN Money. (2004) 'World's most and least expensive cities', Online. Available at http://money.cnn.com/pf/features/popups/costofliving/popup.html (Accessed 20 August 2004).

Coblentz, H. (1963) *Halifax Region Housing Survey*, Halifax, NS, Canada: Central Mortgage and Housing Corporation.

Cohn, M. R. (2004) 'In Shanghai suburbs, bigger's better', *Toronto Star online*, 17 October. Online. Available at www.thestar.com (Archives) (Accessed 17 October 2004).

Collier, R. (1974) *Contemporary Cathedrals*, Montreal: Harvest House.

Congress for the New Urbanism (CNU) (2004a) 'Charter of the new urbanism', Online. Available at http://www.cnu.org/aboutcnu/index.cfm?formaction=charter&CFID=7680873&CFTOKEN=74049180 (Accessed 14 September 2004).

—— (2004b) 'Projects: Karow Nord Master Plan', Online. Available at http://www.cnu.org (Accessed 11 August 2004).

—— (2003) 'Fonti di Matilde, San Bartolomeo, Italy', Online. Available at http://www.cnu.org/about/index.cfm?formAction=project_view&templateInstanceID=678&CFID=7166800&CFTOKEN=7304060 (Accessed 25 April 2003).

Contreras, J. and D. Villano (2002) 'The "new urbanism"', *Newsweek* (10 June) 139(23): 36.

Cooke, P. (1983) *Theories of Planning and Spatial Development*, London: Hutchinson.

Corbett, J. and J. Velasquez (2004) 'The Ahwahnee principles: toward more livable communities', *Local Government Commission*, Online. Available at http://www.lgc.org/freepub/land_use/articles/ahwahnee_article.html (Accessed 7 November 2004).

Council for European Urbanism (CEU) (2004) 'Charter of Stockholm: the charter for European urbanism', Online. Available at http://www.ceunet.org/charter.htm (Accessed 5 August 2004).

—— (2003) 'Draft charter', Online. Available at http://www.ceunet.de/charta.htm (Accessed 8 January 2004).

Council of Europe (2004) 'Heritage', Online. Available at http://www.coe.int/T/E/Cultural_Cooperation/Heritage/Resources/RefTxtCultHer.asp (Accessed 5 August 2004).

Cox, W. (2004) 'Index of articles on urban planning', *Demographia*. Online. Available at www.demographia.com/dbxuplan.htm (Accessed 5 October 2004).

—— (2002) 'Laguna West: new urbanist snout houses', *Demographia*. Online. Available at http://www.demographia.com/dbnucalgw.htm (Accessed 5 October 2004).

Crane, R. (2000) 'Consumer sovereignty and new urbanism'. PLANET, Online posting. Available email: PLANET@LISTSERV.BUFFALO.EDU (Accessed 29 September 2000).

—— (1998) 'Travel by design?', *Access* 12: 2–7.

—— (1996) 'On form versus function: will the new urbanism reduce traffic, or increase it?', *Journal of Planning Education and Research* 15: 117–26.

Creese, W. (1967) *The Legacy of Raymond Unwin*, Cambridge: MIT Press.

—— (1966) *The Search for Environment: The garden city before and after*, New Haven: Yale University Press.

Cusumano, G. (2002) 'Foreword', in C. Bohl, *Place Making*, Washington: Urban Land Institute, pp. viii–xi.

Day, K. (2003) 'New urbanism and the challenges of designing for diversity', *Journal of Planning Education and Research* 23: 83–95.

Denhez, M. (1994) *The Canadian Home: From cave to electronic cocoon*, Toronto: Dundurn Press.

De Roo, G. (2004) 'Coping with the growing complexity of our physical environment: the search for new planning tools in the Netherlands', in A. Sorensen, P. Marcotullio and J. Grant (eds), *Toward Sustainable Cities*, Aldershot: Ashgate, pp. 161–216.

De Roo, G. and D. Miller. (2000) 'Introduction: compact cities and sustainable development', in G. De Roo and D. Miller (eds), *Cities and Sustainable Urban Development: A critical assessment of policies and plans from an international perspective*, Aldershot: Ashgate, pp. 1–13.

Des Moines, City of (2004) 'Vision and mission statement' (Des Moines, Washington) Online. Available at http://www.desmoineswa.gov/city_gov/city_council/vision_mission?vision.html (Accessed 2 November 2004).

DeWolf, C. (2002) 'Why new urbanism fails', *Planetizen*, Online. Available at http://www.planetizen.com/oped/item.php?id=45 (Accessed 27 February 2002).

Dorcey, A. (2004) 'South East False Creek model sustainable community planning process', Online. Available at http://www.interchange.ubc.ca/dorcey/innovations/sefc.html (Accessed 15 October 2004).

Douglas, D., G.-E. Galabuzi, K. Goonewardena, D. Green, J. Hackworth, P. Khosla, S. Kipfer, U. Lehrer, K. Wirsig and D. Young (2004) 'Tearing Regent up', *Now Magazine Online* 26 February – 3 March. Online. Available at http://www.nowtoronto.com/issues/(2004)0226/news_story4.php (Accessed 16 October 2004).

Douglas, M. (1966) *Purity and Danger: An analysis of pollution and taboo*, London: Routledge and Kegan Paul.

Douglas, M. and B. Isherwood (1996) *The World of Goods: Towards an anthropology of consumption*, London: Routledge.

Dover, V. (2004) Presentation in workshop, American Planning Association Conference, Washington DC, 25 April 2004.

Dowling, R. (1998) 'Neotraditionalism in the suburban landscape: Cultural geographies of exclusion in Vancouver, Canada', *Urban Geography* 19(2): 105–22.

Downs, A. (1999) 'Some realities about sprawl and urban decline', *Housing Policy Debate* 10(4): 955–74.

—— (1994) *New Visions for Metropolitan America*, Washington DC: The Brookings Institution and the Lincoln Institute of Land Policy.

Dramstad, W. E., J. D. Olson and R. T. T. Forman (1996) *Landscape Ecology Principles in Landscape Architecture and Land-Use Planning*, Washington: Harvard Graduate School of Design and Island Press.

Duany, A. (2002a) *New Urban Post* IV, Congress for the New Urbanism, p. 6.

—— (2002b) *New Urban Post* IV, Congress for the New Urbanism, p. 7.

—— (2000) 'Gentrification and the paradox of affordable housing'. PRO-URB list, Online posting, Available email PRO-URB@listserv.uga.edu (31 October 2000).

Duany, A. and E. Plater-Zyberk (1994) 'The neighborhood, the district and the corridor', in P. Katz (ed.), *The New Urbanism: Toward an architecture of community*, New York: McGraw-Hill, pp. xvii–xx.

—— (1992) 'The second coming of the American small town', *Wilson Quarterly* 16: 19–50.

Duany, A., E. Plater-Zyberk and R. Alminara (2003) *The New Civic Art*, New York: Rizzoli.

Duany, A., E. Plater-Zyberk and J. Speck (2000) *Suburban Nation: The rise of sprawl and the decline of the American dream*, New York: North Point Press.

Duany, A. and E. Talen (2002a) 'Transect planning', *Journal of the American Planning Association* 68(3): 245–66.

—— (2002b) 'Making the good easy: the smart code alternative', *Fordham Urban Law Journal* 29(4): 1445–68.

Duany Plater-Zyberk Company (2003) *The Lexicon of the New Urbanism*, Version 3.2, Online. Available at www.dpz.com (Accessed 12 September 2004).

Duerkson, C. (2004) Presentation in workshop, American Planning Association Conference, Washington DC, 25 April 2004.

Dumreicher, H., R. Levine, E. Yaranella and T. Radmard (2000) 'Generating models of urban sustainability: Vienna's Westbahnhof sustainable hill town', in K. Williams, E. Burton and M. Jenks (eds), *Achieving Sustainable Urban Form*. London: E&FN Spon, pp. 288–98.

Duncan, A. (2003) 'Expanding accessibility: Vancouver's Mount Pleasant wellness walkways', *Plan Canada* 43(4): 39–41.

Dunlop, B. (1997) 'The new urbanists: the second generation', *Architectural Record* 01.97: 132–4.

Durack, R. (2001) 'Village vices: the contradictions of new urbanism and sustainability', *Places* 14(2): 64–9.

Dyckman, J. (1983) 'Reflections on planning practice in an age of reason', *Journal of Planning Education and Research* 3(1): 5–12.

Ehrenhalt, A. (1997) 'The dilemma of the new urbanists', (Originally in *Governing Magazine*) Online. Available at www.state.fl.us/fdi/news/state/urbnst.htm (Accessed 16 February 2004).

Ellis, C. (2002) 'The new urbanism: critiques and rebuttals', *Journal of Urban Design* 7(3): 261–91.

Eppli, M. J. and C. C. Tu (1999) *Valuing the New Urbanism: Impact of the new urbanism on prices of single-family homes*. Washington DC: Urban Land Institute.

Eurocouncil (2004) 'Purpose and agenda', Online. Available at http://www.eurocouncil.net/id17.htm (Accessed 10 April 2004).

—— (2003) 'Euro new urbanism', Online. Available at http://www.eurocouncil.net/id21_m.htm (Accessed 5 October 2003).

European Environment Agency (2001) 'Indicator: vehicle ownership', Online. Available at http://themes.eea.eu.int/Sectors_and_activities/transport/indicators/spatial/transport/Vehicle_ownership_TERM_(2001)doc.pdf (Accessed 5 August 2004).

Everett-Green, R. (1997) 'Model cities for the next millennium', *The Globe and Mail* (Toronto), 31 May, C21.

Ewing, R. (1996) *Best Development Practices: Doing the right thing and making money at the same time*, Chicago: Planners Press.

Fader, S. (2000) 'Density by design', *Urban Land* 59(7): 55–9.

Fainstein, S. (2000) 'New directions in planning theory', *Urban Affairs Review* 35(4): 451–78.

Falconer Al-Hindi- *see Al-Hindi*.

Filion, P. 'Smart growth in theory and practice', Public presentation, Dalhousie University, Halifax NS, 13 April 2004.

—— (1996) 'Metropolitan planning objectives and implementation constraints: planning in a post-Fordist and postmodern age', *Environment and Planning A* 28(9): 1637–60.

Fischler, R. (2004) Book review of J. Hillier, 'Shadows of power: an allegory of prudence in land-use planning', Routledge, *Journal of the American Planning Association* 70(3): 363.

Fishman, R. (1977) *Urban Utopias in the Twentieth Century: Ebenezer Howard, Frank Lloyd Wright, and Le Corbusier*, New York: Basic Books.

Fletcher Farr Ayotte. (2001) 'Orenco Station', Online. Available at http://www.ffadesign.com/features/orenco.htm (Accessed 7 October 2004).

Flyvberg, B. (2001) *Making Social Science Matter: Why social inquiry fails and how it can succeed again*, Cambridge: Cambridge University Press.

Foglesong, R. (1986) *Planning the Capitalist City: The colonial era to the 1920s*, Princeton NJ: Princeton University Press.

Ford, L. (2001) 'Alleys and urban form: testing the tenets of new urbanism', *Urban Geography* 22(3): 268–86.

Forester, J. (1989) *Planning in the Face of Power*, Berkeley: University of California Press.

Forman, R.T.T. and M. Godron (1986) *Landscape Ecology*, New York: John Wiley and Sons.

Foucault, M. (1977) *Discipline and Punish: The birth of the prison*, trans. A. Sheridan, New York: Vintage Books Random House.

Frank, R. (1999) *Luxury Fever: Why money fails to satisfy in an era of excess*, New York: Free Press.

Franklin, B. (2003) 'Success or failure? The redevelopment of Bordesley as an (urban) village', *Urban Design International* 8: 21–35.

Franklin, B. and M. Tait (2002) 'Constructing an image: the urban village concept in the UK', *Planning Theory* 1(3): 250–72.

Frantz, D. and C. Collins (1999) *Celebration, U.S.A: Living in Disney's brave new town*, New York: Henry Holt and Company.

Frey, H. (1999) *Designing the City*, London: E & FN Spon.

Friedman, A. (2002) *Planning the New Suburbia: Flexibility by design*, Vancouver: University of British Columbia Press.

Friedmann, J. (1987) *Planning in the Public Domain: From knowledge to action*, Princeton NJ: Princeton University Press.

—— (1979) *The Good Society*, Cambridge, MA: MIT Press.

Friedmann, J., and B. Hudson (1974) 'Knowledge and action: a guide to planning theory', *Journal of the American Institute of Planners* 40(1): 2–16.

Frumkin, H., L. Frank and R. Jackson (2004) *Urban Sprawl and Public Health: Designing, planning, and building for healthy communities*, Washington DC: Island Press.

Fujita, K. and R. Child Hill (1997) 'Together and equal: place stratification in Osaka', in P.P. Karan and K. Stapleton (eds), *The Japanese City*, Lexington: University Press of Kentucky, pp. 106–33.

Fulton, W. (1996) *The New Urbanism: Hope or hype for American communities?* Cambridge MA: Lincoln Institute of Land Policy.

Fung, A. (2001) 'Beyond and below the new urbanism: citizen participation and responsive spatial reconstruction', *Boston College Environmental Affairs Law Review* 28(4): 615–35.

Garde, A. M. (2004) 'New urbanism as sustainable growth? a supply side story and its implications for public policy', *Journal of Planning Education and Research* 24: 154–70.

GarrisonWoods.net. (2000) 'Neighbourhood update'. Garrison Woods Newsletter, April, Online. Available at http://www.garrisonwoods.net/articles/newsletter01.html (Accessed 31 May 2002).

Gause, J. Allen (2002) *Great Planned Communities*, Washington: Urban Land Institute.

Gehl, J. and L. Gemzoe (2003) *New City Spaces*, Copenhagen: The Danish Architectural Press.

GHK Canada (2002) *Growing Together: Proposals for renewal in the Toronto region*, City of Toronto Planning Department.

Giddens, A. (1991) *Modernity and Self-Identity: Self and society in the late modern age*, Stanford, CA: Stanford University Press.

Gindroz, R. (2000) 'Chapter twenty one', in M. Leccese and K. McCormick (eds), *Charter of the New Urbanism*, Congress for the New Urbanism. New York: McGraw Hill.

Giroir, G. (2003) 'Gated communities, clubs in a club system: The case of Beijing (China)', Paper presented at the conference, *Gated communities: building safer communities or social division?*, Glasgow, September. Online. Available at http://www.bristol.ac.uk/sps/cnrpapersword/gated/giroir.doc (Accessed 20 August 2004).

Girouard, M. (1985) *Cities and People*, New Haven, CT: Yale University Press.

Goodman, W. M. and E. C. Freund (1968) *Principles and Practice of Urban Planning*, Washington DC: International City Managers Association.

Goetz, E. G. (2003) *Clearing the Way: Deconcentrating the poor in urban America*, Washington DC: The Urban Institute Press.

Gordon, D. and S. Fong (1989) 'Designing St. Lawrence'. In D. Gordon (ed.), *Learning from St. Lawrence*, Conference Proceedings, Toronto: Ryerson Polytechnical Institute.

Gordon, D. and K. Tamminga (2002) 'Large-scale traditional neighbourhood development and pre-emptive ecosystem planning: the Markham experience 1989–2001', *Journal of Urban Design* 7(2): 321–40.

Gordon, P. and H. Richardson (1997) 'Are compact cities a desirable planning goal?', *Journal of the American Planning Association* 63(1): 95–106.

Graham, S. and P. Healey (1999) 'Relational concepts of space and place: issues for planning theory and practice', *European Planning Studies* 7(5): 623–46.

Grant, J. (2004) 'Sustainable urbanism in historical perspective', in A. Sorensen, P. Marcotullio and J. Grant (eds), *Towards Sustainable Cities: East Asian, North American, and European Perspectives*, Aldershot: Ashgate, pp. 24–37.

—— (2003) 'Exploring the influence of new urbanism', *Journal of Architectural and Planning Research* 20(3): 234–53.

—— (2002a) 'From "sugar cookies" to "gingerbread men": conformity in suburban design', *Planners Network* 151: 10–3. Also available online at http://www.plannersnetwork.org/htm/pub/archives/152/grant.htm (Accessed 17 October 2004).

Grant, J. (2002b) 'Mixed use in theory and practice: Canadian experience with implementing a planning principle', *Journal of the American Planning Association* 68(1): 71–84.
—— (2001a) 'The dark side of the grid: power and urban design', *Planning Perspectives* 16: 1–23.
—— (2001b) 'Sustainable communities: lessons from Japan', *Plan Canada* 41(2): 9–11.
—— (2000a) 'Life with less sprawl: Japanese communities demonstrate designs for more compact living', *Alternatives* 26(3): 24.
—— (2000b) 'Planning Canadian cities: context, continuity, and change', Chapter 19 in T. Bunting and P. Filion (eds), *The Canadian City in Transition* (2nd edn), Toronto: Oxford University Press, pp. 443–61.
—— (1994) *The Drama of Democracy: Contention and dispute in community planning*, Toronto: University of Toronto Press.
—— (1990). 'Understanding the social context of planning', *Environments* 20(3): 10–9.
Grant, J., P. Manuel and D. Joudrey (1996) 'A framework for planning sustainable residential landscapes', *Journal of the American Planning Association* 62(3): 331–44.
Grant, J., P. Marcotullio and A. Sorensen (2004) 'Towards land management policies for more sustainable cities', in A. Sorensen, P. Marcotullio and J. Grant (eds), *Towards Sustainable Cities: East Asian, North American and European Perspectives on Managing Urban Regions*, Aldershot: Ashgate, pp. 301–8.
Grant, J., H. Saito and K. Itoh (2002) 'The effects of administrative structure on planning outcomes: comparing Canada and Japan', *Faculty of Sociology Bulletin* (Chukyo University, Toyota, Japan) 16(1): 87–107.
Greenwald, M. J. (2003) 'The road less traveled: New Urbanist inducements to travel mode substitution for nonwork trips', *Journal of Planning Education and Research* 23: 39–57.
Greenwich Millennium Village (2004) 'Greenwich Millennium Village'. Online. Available at http://www.greenwich-village.co.uk (Accessed 6 August 2004).
Gross, B. (1980) *Friendly Fascism: The new face of power in America*, Montreal: Black Rose Books.
Guelph, City of (2004) *Smart Guelph: Building tomorrow today* (Report to Council) Guelph.
Gyourko, J. E. and W. Rybczynski (2004) 'Financing new urbanism', *Urban Land* 63(5): 18–25.
—— (2000) 'Financing new urbanism projects: obstacles and solutions', *Housing Policy Debate* 11(3): 733–50.
Hague, C. (2001) 'One city, two identities', *Regeneration and Renewal* (6 April): 16–18.
—— (1997) 'Defining a role for the urban village', *Planning* 1208 (7 March): 13.
Halifax Regional Municipality (HRM) (2001) 'Healthy Growth for HRM. Why regional planning?', Online. Available at http://wysiwyg://61/http://www.region.halifax.ns.ca/regionalplanning/index.html (Accessed 5 July 2002).

Hall, D. D. (1998) 'Community in the new urbanism: design vision and symbolic crusades', *Traditional Dwellings and Settlement Review* 9(11): 23–36.

Hall, J. W. (1970) *Japan: From prehistory to modern times*, Tokyo: Charles E. Tuttle.

Hall, P. (1988) *Cities of Tomorrow: An intellectual history of urban planning and design in the twentieth century*, Oxford: Basil Blackwell.

Hammond, M. (1972) *The City in the Ancient World*, Cambridge, MA: Harvard University Press.

Hancock, M. (1994) 'Don Mills, a paradigm of community design', *Plan Canada* 34 (special 75th anniversary edition, July): 87–90.

Handwerk, B. (2004) 'China's car boom tests safety, pollution practices', *National Geographic News*. Online. Available at http://news.nationalgeographic.com/news/(2004)/06/0628_040628_chinacars.html#main (Accessed 20 August 2004).

Handy, S. L. (1996) 'Urban form and pedestrian choices: study of Austin neighborhoods', *Transportation Research Record* 1552: 135–44.

—— (1992) 'Regional versus local accessibility: neo-traditional development and its implications for non-work travel', *Built Environment* 18: 253–67.

Harrison, M. (1999) *Bournville: Model village to garden suburb*, Chichester, West Sussex: Phillimore and Company.

Hardy, D. (2003) 'Poundbury – planning by royal appointment', *Town and Country Planning* (June): 155–7.

Hardoy, J. (1968) *Urban Planning in Pre-Columbian America*, London: Studio Vista.

Harvey, D. (1997) 'The new urbanism and the communitarian trap', *Harvard Design Magazine* 1: 1–3, Online. Available at http:mitpress.mit.edu/HDM (Accessed 10 September 2003).

—— (1994) 'Flexible accumulation through urbanization: reflections on "Post-modernism" in the American city', in A. Amin (ed.), *Post-Fordism: A reader*, Oxford: Blackwell, pp. 361–86.

—— (1989) *The Urban Experience*, Baltimore: Johns Hopkins University Press.

—— (1973) *Social Justice and the City*, Baltimore: Johns Hopkins University Press.

Haworth, L. (1963) *The Good City*, Bloomington, IN: Indiana University Press.

Hayden, D. (2000) 'Model houses for the millions: the making of the American suburban landscape, 1820-(2000)', *Working Paper WP00DHZ*, Cambridge, MA: Lincoln Institute of Land Policy.

—— (1995) *The Power of Place: Urban landscapes as public history*, Cambridge MA: MIT Press.

Healey, P. (1997) *Collaborative Planning: Shaping places in fragmented societies*, Vancouver: UBC Press.

Health Canada. (2002) *Ottawa Charter for Health Promotion*. An International Conference on Health Promotion, Online. Available at http://www.hcsc.gc.ca/hppb/phdd/docs/charter/ (Accessed 6 May 2004).

Hebbert, M. (1986) 'Urban sprawl and urban planning in Japan', *Town Planning Review* 57(2): 141–58.

Heikkila, E. J. and M. Griffin (1995) 'Confucian planning or planning confusion?', *Journal of Planning Education and Research* 14: 269–79.

Hendler, S. (1989) 'The Canadian Healthy Communities Project: relevant or redundant?', *Plan Canada* 29(4): 32–4.

Higgins, D. (1986) *Local and Urban Politics in Canada*, Toronto: Gage Educational Publishing.

Higgs, E., L. Martin and P. Filion (1988) 'A survey of the preferred literature of Canadian planners', *Plan Canada* 28(1): 6–16.

Hodge, G. (2003) *Planning Canadian Communities: An introduction to the principles, practice and participants* (4th edn), Toronto: Thomson Nelson.

Horne, D. (1994) *The Public Culture: An argument with the future* (2nd edn), London: Pluto Press.

Houstoun, F. (2004) 'Growth consensus: smart growth stalls when it polarizes rather than leads', *Urban Land* 63(6): 14–16.

Howard, E. (1985 [1902]) *Garden Cities of To-morrow*, Eastbourne, UK: Attic Books (New illustrated reprinted edition).

Hubbard, E. and M. Shippobottom (1988) *A Guide to Port Sunlight Village*, Liverpool: Liverpool University Press.

Hudson, B. (1979) 'Comparison of current planning theories', *Journal of the American Planning Association* 45(2): 387–96.

Hughes, D. J. (1994) *Pan's Travail: Environmental problems of the Ancient Greeks and Romans*, Oxford: Blackwell.

Huppert, G. (1986) *After the Black Death: A social history of early modern Europe*, Bloomington, IN: Indiana University Press.

Hygeia Consulting Services and REIC Ltd (1995) *Changing Values, Changing Communities: A guide to the development of healthy, sustainable communities*, Ottawa: Canada Mortgage and Housing Corporation.

Innes de Neufville, J. (1983) 'Planning theory and practice: bridging the gap', *Journal of Planning Education and Research* 3(1): 35–45.

Iovine, J.V. (1996) 'Our town? The nostalgic new urbanism is running into trouble in the real world', *Utne Reader* 73 (Jan./Feb.): 30.

Ishikawa, M. (2004) 'Green structure plan for a sustainable urban–rural relationship', in A. Sorensen, P. Marcotullio and J. Grant (eds), *Towards Sustainable Cities: East Asian, North American and European Perspectives on Managing Urban Regions*, Aldershot: Ashgate, pp. 228–38.

Isin, E. and R. Tomalty (1993) *Resettling Cities: Canadian residential intensification initiatives*, Ottawa: Canada Mortgage and Housing Corporation.

Jacobs, J. (1961) *Death and Life of Great American Cities*, New York: Vintage Books.

James, S. (2003) 'Blooming boulevards: a new tool for green space management', *Plan Canada* 43(4): 37–8.

Jenks, M., E. Burton and K. Williams (eds) (1996) *The Compact City: A sustainable urban form?* London: E & FN Spon.

Kaplan, R. D. (1998) *An Empire Wilderness: Travels into America's future*, Toronto: Random House of Canada.

Kaplan, S. H. (1990) 'The holy grid: a skeptic's view', *Planning* 56(11): 10–11.

Karan, P. P. (1997) 'The city in Japan', in P. P. Karan and K. Stapleton (eds), *The Japanese City*. Lexington, Kentucky: University Press of Kentucky, pp. 12–39.

Katz, P. (2002) 'Individual investors can profit from new urbanism', *New Urban News* 7(6): 12–14.

—— (ed.) (1994) *The New Urbanism: Toward an architecture of community*, New York: McGraw Hill.

Kay, J. Holtz (1997) *Asphalt Nation: How the automobile took over America, and how we can take it back*, New York: Crown Publishers.

Kelbaugh, D. (2002) *Repairing the American Metropolis: Common place revisited*, Seattle: University of Washington Press.

—— (1997) *Common Places: Toward neighborhood and regional design*, Seattle: University of Washington Press.

Kelbaugh, D. (ed.) (1989) *The Pedestrian Pocket Book : A new suburban design strategy*, New York: Princeton Architectural Press.

Kellett, J. R. (1969) *The Impact of Railways on Victorian Cities*, London: Routledge and Kegan Paul.

Kelly, B. (1993) *Expanding the American Dream: Building and rebuilding Levittown*, Albany, NY: State University of New York Press.

Kim, J. and R. Kaplan (2004) 'Physical and psychological factors in sense of community: new urbanist Kentlands and nearby Orchard Village', *Environment and Behavior* 36(3): 313–40.

Kim, K.-J. (2004) 'Inner-city growth management problem in Seoul: residential rebuilding boom and planning response', in A. Sorensen, P. Marcotullio, and J. Grant (eds), *Towards Sustainable Cities: East Asian, North American and European perspectives on managing urban regions*, Aldershot UK: Ashgate, pp. 267–84.

Klosterman, R. (1980) 'A public interest criterion', *Journal of the American Planning Association* 46: 323–33.

Knowles, P. (2003) 'Designing out crime: the cost of policing new urbanism', Operation Scorpion web site, Bedfordshire Police, UK. Online. Available at http://www.operationscorpion.org.uk/design_out_crime/policing_urbanism.htm (Accessed 5 June 2004).

Knox, P. (1993) 'Capital, material culture and socio-spatial differentiation', in P. Knox (ed.), *The Restless Urban Landscape*, Englewood Cliffs, NJ: Prentice Hall, pp. 1–34.

Knox, P. (1992) 'The packaged landscapes of post-suburban America', in J. W. R. Whitehand and P.J. Larkham (eds), *Urban Landscapes: International perspectives*, London: Routledge, pp. 207–26.

Kobayashi, G. (2000) 'Buildings in Japan: built today, demolished soon', *Daily Yomiuri Online*, 13 November. www.yomiuri.co.jp/report-e/rep03_(2000)a.htm (Accessed 14 November 2000).

Koizumi, H. (2004) 'Empowerment in the Japanese planning context', in A. Sorensen, P. Marcotullio and J. Grant (eds), *Towards Sustainable Cities: East Asian, North American and European Perspectives on managing urban regions*, Aldershot: Ashgate, pp. 217–27.

Koskela, K. (2000) '"The gaze without eyes": video-surveillance and the changing nature of urban space', *Progress in Human Geography* 24(2): 243–65.

Kreyling, C. (2001) 'Fat city', *Planning*, 67(6): 4–9.

Krieger, A. (1998) 'Whose urbanism?', *Architecture* 87(11): 73–7.

—— (ed.) (1991) *Andres Duany and Elizabeth Plater-Zyberk: Towns and town-making principles*, Harvard University Graduate School of Design, New York: Rizzoli.

Krier, L. (2001) 'Interview with Nikos Salingaros'. *Planetizen*, 5 November. Online. Available at www.planetizen.com/oped/item.php?id=35 (Accessed 8 January 2004).

—— (1998) *Architecture: Choice or fate*, Singapore: Andreas Papadakis Publisher.

—— (1991) 'Afterword', in A. Krieger (ed.), *Andres Duany and Elizabeth Plater-Zyberk: Towns and town-making principles*, New York: Rizzoli, pp. 117–9.

—— (1984a) 'Drawings' (Originally in Archives d'Architecture Moderne, Brussels, xxv–xxxi, 1980), Revised version in *Architectural Design* 54 (Nov/Dec), pp. 16–22.

—— (1984b) *Houses, Palaces, Cities*, D. Porphyrios (ed.), London: Architectural Design Editions.

—— (1978) 'The reconstruction of the city', *Rational Architecture. La reconstruction de la ville Europeanne*, Bruxelles: Archives d'Architecture Moderne, pp. 38–42, 163–80.

—— (1977) 'The city within the city', *A&U Tokyo, Special issue*, November, pp. 69–152, Online. Available at http:// applied.math.utsa.edu/krier/city.html (Accessed 8 January 2004).

Kunstler, J. H. (1999) 'Where evil dwells: reflections on the Columbine School massacre', Address given at the Congress for the New Urbanism, Milwaukee, 6 June 1999, Online. Available at www.kunstler.com/spch_milw.html (Accessed 27 May 2004).

—— (1996) *Home from Nowhere: Remaking our every day world for the twenty-first century*, New York: Simon and Schuster.

—— (1993) *The Geography of Nowhere: The rise and decline of America's man-made landscape*, New York: Simon and Schuster.

Landman, K. (2003) 'Sustainable "urban village" concept: mandate, matrix or myth?', Online. Available at www.csir.co.za/akani/(2003)/jul/07_landman.pdf (Accessed 9 February 2004).

Langdon, P. (2004a) 'Transportation establishment is warming to new urbanist ideas', *New Urban News* 9(1): 1, 6–8.
—— (2004b) 'Tradition-minded architects aim to expand their influence', *New Urban News* 9(4): 11–12, 14.
—— (2004c) 'Three years after 9/11, security mindset threatens civic design', *New Urban News* 9(6): 1, 3–5.
—— (2004d) 'Poundbury thrives despite a protest against density', *New Urban News* 9(7): 5–6.
—— (2003a) 'Zoning reform advances against sprawl and inertia', *New Urban News* 8(1): 1, 3–5.
—— (2003b) 'As retail chains grow, threats to local character increase', *New Urban News* 8(8): 1, 3.
—— (2003c) 'In central Vancouver, modernism and new urbanism mesh', *New Urban News* 8(8): 8–10.
—— (2002a) 'The visitability challenge', *New Urban News* 7(6): 1, 4–6.
—— (2002b) 'Critics collide over "urban network" proposal', *New Urban News* 7(7): 1, 5, 7.
Lasch, C. (1979) *The Culture of Narcissism: American life in an age of diminishing expectations*, New York: Norton.
Lassar, T. (2001) 'Destiny with density', *Urban Land* 60(3): 43–7.
Law Development Group (1996) '1,500-acre housing project set for Toronto area (Cornell)', *Financial Post Daily* 9(144): 5.
Leccese, M. and K. McCormick (eds) (2000) *Charter of the New Urbanism*, Congress for the New Urbanism, New York: McGraw Hill.
Lee, S.-D. (2004) 'Urban growth management and housing supply in the Capital Region of South Korea', in A. Sorensen, P. Marcotullio and J. Grant (eds), *Towards Sustainable Cities: East Asian, North American and European Perspectives on managing urban regions*, Aldershot: Ashgate, pp. 285–98.
Leinberger, C. (1993) *Suburbia: Land Use in Transition*, Washington: Urban Land Institute.
Lennard, S. Crowhurst and J. Riley Jr. (2004) 'Children and the built environment', *Urban Land* 63(1): 68–9.
Lennertz, W. (2000) 'Chapter seventeen', in M. Leccese and K. McCormick (eds), *Charter of the New Urbanism*, Congress for the New Urbanism, New York: McGraw Hill, pp. 109–12.
—— (1991) 'Town-making fundamentals', in A. Krieger (ed.), *Andres Duany and Elizabeth Plater-Zyberk: Towns and town-making principles*, Harvard University Graduate School of Design, New York: Rizzoli, pp. 21–4.
Leung, H.-L. (1995) 'A new kind of sprawl', *Plan Canada* 35(5): 4–5.
Levine, L. (1988) *Highbrow/Lowbrow: The emergence of cultural hierarchy in America*, Cambridge, MA: Harvard University Press.

Lianos, M. (2003) 'Social control after Foucault', *Surveillance and Society* 1(3): 412–30. Online. Available at http://www.surveillance-and-society.org/articles1(3)/After Foucault.pdf (Accessed 29 May 2004).

Liberal Party of Canada (2002) 'Prime Minister's Caucus Task Force on Urban Issues releases final report', Online. Available at http://www.liberal.parl.gc.ca/urb/home_e.htm (Accessed 16 October 2004).

Local Government Commission (2002) 'Ahwahnee principles', Online. Available at http://www.lgc.org/ahwahnee/principles.html (Accessed 7 November 2004).

Lorinc, J. (2001) 'The story of sprawl', *Toronto Life* 35(8): 82–100.

Lund, H. (2003) 'Testing the claims of new urbanism: local access, pedestrian travel, and neighboring behaviors', *Journal of the American Planning Association* 69(4): 414–29.

—— (2002) 'Pedestrian environments and sense of community', *Journal of Planning Education and Research* 21: 301–12.

Lynch, K. (1981) *A Theory of Good City Form*, Cambridge, MA: MIT Press.

MacDonald, D. and B. Clark (1995) 'New urbanism in Calgary: McKenzie Towne', *Plan Canada* 35(4): 20–2.

Machiavelli, N. (1950[1513]) *The Prince and the Discourses*, New York: The Modern Library.

Mallet, M. (2004) 'Great expectations', The Vancouver Courier.com, 19 January, Online. Available at http://www.vancourier.com/014104/news/014104nn1.html (Accessed 28 January 2004).

Manzi, T. and B. Smith-Bowers. (2003) 'Gated communities and mixed tenure estates: segregation or social cohesion?', Paper presented at conference "Gated communities: building social division or safer communities?", Glasgow, September 18–19, 2003, Online. Available at: http://www.bristol.ac.uk/sps/cnrpapersword/gated/manzi.pdf (Accessed 4 February 2004).

Marcotullio, P. (2004) 'Why the Asian urbanization experience should make us think differently about planning approaches', in A. Sorensen, P. Marcotullio and J. Grant (eds), *Towards Sustainable Cities: East Asian, North American and European Perspectives on managing urban regions*, Aldershot: Ashgate, pp. 38–58.

Marcuse, P. (2000) 'The new urbanism: the dangers so far', *DISP* 140: 4–6. Online. Available at http://disp.ethz.ch/pdf/140_1.pdf (Accessed 5 December 2003).

Markham, Town of (2003) *Cornell Community in the Town of Markham: Architectural design guidelines*, Toronto: Prepared by Watchorn Architect.

Markham, Town of (1998) *Design Implementation Guidelines*, Markham ON: Development Services Commission.

—— (1991) *The Markham Village Heritage Conservation District*, Markham, ON: Development Services Commission.

Marshall, A. (2004) 'Suburbs in disguise', Online. Available at http://www.alexmarshall.org/index.php?pageId=75 (Accessed 27 October 2004).

—— (2000) *How Cities Work: Suburbs, sprawl, and the roads not taken*, Austin: University of Texas Press.

—— (1995a) 'The demolition man', *Metropolis* (May) Online. Available at www.alexmarshall.org/index.htm?articleId=53 (Accessed 16 February 2004).

—— (1995b) 'What makes a neighborhood viable?', *Metropolis* (May) Online. Available at www.alexmarshall.org/index.htm?articleId=54 (Accessed 16 February 2004).

—— (1995c) 'When the new urbanism meets an old neighborhood', *Metropolis* (May) Online. Available at www.alexmarshall.org/am_articleFolder/new_urbanism_vs_old_neigh.htm (Accessed 5 October 2004).

Martin, C. (1989) 'Second chance', in A. Papadakis (ed.), *Prince Charles and the Architectural Debate*. Architectural Design, London: St Martin's Press, pp. 7–15.

Martin, J. (1996) 'Building "community"', *Urban Land* 55(3): 28–32.

Mather, C. (1997) 'Urban landscapes of Japan', in P. P. Karan and K. Stapleton (eds), *The Japanese City*, Lexington: University Press of Kentucky, pp. 40–55.

McCann, E. (1995) 'Neotraditional developments: the anatomy of a new urban form', *Urban Geography* 16(3): 210–33.

McHarg, I. (1969) *Design with Nature*, Garden City, New York: American Museum of Natural History, Natural History Press.

McInnes, C. (1992) 'Drawing happiness into the blueprints', *The Globe and Mail* (Toronto), 27 April, A17.

McKenzie, E. (1994) *Privatopia: Homeowner associations and the rise of residential private government*, New Haven, CT: Yale University Press.

Mendez, M. (2005) 'Latino new urbanism: building on cultural preferences', *Opolis* 1(1): 33–48 Online. Available at http://www.mi.vt.edu/uploads/Opolis/Mendez.pdf (Accessed 4 December 2004).

Metropolitan Toronto, Municipality of (1987) *Housing Intensification*, Metropolitan Plan Review. Toronto: Policy Development Division, Metropolitan Toronto Planning Department.

Metropolitan Transportation Commission (MTC), San Francisco Bay Area (2004) 'Table A-3 International auto and vehicle ownership comparison – 1995', Online. Available at http://www.mtc.ca.gov/datamart/forecast/ao/tablea3.htm (Accessed 1 October 2004).

Metropolis. (1995) 'What makes a neighborhood viable? A roundtable debate', Alex Marshall and Andres Duany, May. www.alexmarshall.org/index.htm?articleId=54 (Accessed 16 February 2004).

Milgrom, R. (2002) 'Engaging new urbanism', *Planners Network* 151: 2, 9.

Miller, M. (1989) *Letchworth: The first garden city*, Chichester: Phillimore and Co.

Mitchell, J. G. (2001) 'Urban sprawl', *National Geographic* (July) 200(1): 48–73.

Mohney, D. (1991) 'Preface and acknowledgements', in D. Mohney and K. Easterling (eds), *Seaside: Making a town in America*, New York, Princeton Architectural Press, pp. 36–47.

Mohney, D. and K. Easterling (1991) *Seaside: Making a town in America*, New York: Princeton Architectural Press.

Morris, A. E . J. (1994) *History of Urban Form Before the Industrial Revolutions*, Harlow, UK: Longman Scientific.

Morris, D. (1969) *The Human Zoo*, London: Cape.

—— (1968) *The Naked Ape: A zoologist's study of the human animal*, London: Cape.

Morris, I. (1964) *The World of the Shining Prince: Court life in ancient Japan*, New York: Penguin Books.

Moule, E. and S. Polyzoides (1994) 'The street, the block, and the building', in P. Katz (ed.), *The New Urbanism: Toward an architecture of community*, New York: McGraw-Hill, pp. xxi–xxiv.

MSM Regional Council and the Regional Plan Association (MSM/RPA) (1994) *Redesigning the Suburbs: Turning sprawl into centers*, Middlesex Somerset Mercer Counties, New Jersey.

Mumford, L. (1961) *The City in History*, New York: Harcourt, Brace and World.

Myers, D. (2000) 'Consumer sovereignty and new urbanism', PLANET, Online posting. Available email: PLANET@LISTSERV.BUFFALO.EDU (29 September 2000).

Nara Museum. (n.d.) *Nara Imperial Palace Site Museum*, Pamphlet, Nara, Japan.

Nasar, J. (2003) 'Does neotraditional development build community?', *Journal of Planning Education and Research* 23: 58–68.

Neuman, J. (2003) 'Building in affordability', *Urban Land* 62(5): 64–5.

New Urban News (2004a) 'Codes make a difference in California, Virginia', (July/Aug) 9(5): 6.

—— (2004b) 'Small lots win residents' approval if parks are nearby, survey finds', (Sept) 9(6): 10–11.

—— (2003a) 'Empty Providence buildings fill up again', (Oct/Nov) 8(7): 11–12.

—— (2003b) 'New urbanism adds to housing value, study says', (Dec) 8(8): 4.

—— (2002) 'State planning reforms make gains' (Apr/May) 7(3): 1, 4–5.

Newsweek. (1995) 'Paved paradise: 15 ways to fix the suburbs' (15 May) 125(20): 40–53.

Nitobe, I. (1969) *Bushido: The soul of Japan*, Tokyo: Tuttle.

Nova Scotia Round Table on Environment and Economy (1992) *The Sustainable Development Strategy for Nova Scotia*, Halifax: Government of Nova Scotia Printer.

Office of the Deputy Prime Minister (ODPM) (2004a) 'Creating sustainable communities: Planning policy guidance notes', Online. Available at http://www.odpm.gov.uk/stellent/groups/odpm_control/documents/contentservertemplate/odpm_index.hcst?n=2263&l=2 (Accessed 24 June 2004).

—— (2004b) 'Safer places: the planning system and crime prevention', Online. Available at http://www.odpm.gov.uk/stellent/groups/odpm_planning/documents/page/odpm_plan_028449.pdf (Accessed 6 August 2004).

—— (1999) 'Millennium villages and sustainable communities', (LlewelynDavies, CAG Consultants and GHK Economics), Online. Available at http://www.odpm.gov.uk/stellent/groups/odpm_urbanpolicy/documents/page/odpm_urbpol_60809602.hcsp (Accessed 12 August 2004).

Okamoto, K. (1997) 'Suburbanization of Tokyo and the daily lives of suburban people', in P. P. Karan and K. Stapleton (eds), *The Japanese City*, Lexington: University Press of Kentucky, pp. 79–105.

O'Keefe, K. (2002) 'Peter Katz: marketing the new urbanism', *The Town Paper* 4(4) Online. Available at http://www.tndtownpaper.com/Volume4/peter_katz.htm (Accessed 13 September 2004).

Olson, S. (2002) 'Vancouver is transformed into a city of glass', *Architectural Record* 190(4): 55–6.

Ontario, Government of (1997) *Breaking Ground: An illustration of alternative development standards in Ontario's new communities*, Toronto: Queen's Printer.

—— (1995) *Alternative Development Standards: Making choices*, Toronto: Queen's Printer.

O'Toole, R. (2004) 'Biographic web page'. Online. Available at http://www.urbanfutures.org/otoole.html (Accessed 5 October 2004).

—— (2003) 'New urbanism promotes crime', Reason Public Policy Institute. Online. Available at http://www.rppi.org/newurbancrime.shtml (Accessed 6 August 2004).

Ouellet, M. (2004) 'The relationship between the American-based new urbanism movement and contemporary neotraditional urban design in Europe', unpublished thesis, Dalhousie University, Halifax, Canada.

Owens, E. J. (1991) *The City in the Greek and Roman World*, London: Routledge.

Ozaki, R. (2002) 'Housing as a reflection of culture: privatised living and privacy in England and Japan', *Housing Studies* 17(2): 209–27.

Pateman, M. L. (2004) 'What is new urbanism?', *Urban Land* 63(7): 17.

Peck and Associates (2000) *Implementing Sustainable Community Development: Charting a federal role for the 21st century*, Ottawa: Canada Mortgage and Housing Corporation.

Perin, C. (1977) *Everything in its Place: Social order and land use in America*, Princeton NJ: Princeton University Press.

Perkins, B. (2001) 'Special report: Santana Row, suburban urban oasis', Deadline News, Online. Available at http://www.deadlinenews.com/santanarow063001.html (Accessed 10 April 2004).

Perry, C. (1974 [1929]) 'The neighborhood unit', *Monograph 1, Neighborhood and*

Community Planning, Regional Survey of New York and Environs, Volume VII. (Reprinted, New York: Arno Press).
Pirenne, H. (1956 [1925]) *Medieval Cities: Their origins and the revival of trade*, Garden City NY: Doubleday Anchor Books reprint.
Plaut, P.O. and M. Boarnet (2003) New urbanism and the value of neighborhood design. *Journal of Architectural and Planning Research* 20(3): 254–65.
Pogharian, S. (1996) 'Street design: learning from suburbia', *Plan Canada* 36(5): 41–2.
Popkin, S., B. Katz, M. K. Cunningham, K. Brown, J. Gustafson and M. Austin Turner. (2004) *A Decade of HOPE VI: research findings and policy challenges*, The Urban Institute, Online. Available at http://www.urban.org/urlprint.cfm?ID=8864 (Accessed 5 September 2004).
Prince of Wales, Charles (2004) 'Speeches and articles: architecture' (Web site of the Prince of Wales), Online. Available at http://www.princeofwales.gov.uk/speeches/speeches_ index_arc.html (Accessed 5 August 2004).
—— (1989) *A Vision of Britain*, London: Doubleday.
Putnam, R. (1995) 'Bowling alone: America's declining social capital', *Journal of Democracy* 6(1): 65–78.
Pyatok, M. (2002) 'The narrow base of the new urbanists', *Planners Network* 151: 1, 4–5. Online. Available at http://www.plannersnetwork.org/htm/pub/archives/152/pyatok.htm (Accessed 16 October (2004).
—— (2000) 'Comment on Charles C. Bohl's "New urbanism and the city". The politics of design: the new urbanists vs. the grass roots', *Housing Policy Debate* 11(4): 803–14.
Rabinow, P. (1984) *The Foucault Reader*, New York: Pantheon Books.
Rae, D. (2003) *City: Urbanism and its end*, New Haven: Yale University Press.
Rees, A. (2003) 'New urbanism: visionary landscapes in the twenty-first century', in M. Lindstrom and H. Bartling (eds), *Suburban Sprawl: Culture, theory, and politics*, Oxford: Rowman and Littlefield, pp. 93–114.
Relph, E. (1987) *The Modern Urban Landscape*, Baltimore: Johns Hopkins University Press.
Richardson, N. (1989) *Land Use Planning and Sustainable Development in Canada*, Ottawa: Canadian Environmental Advisory Agency.
Riger, S. and P. J. Lavrakas (1981) 'Community ties: patterns of attachment and social interaction in urban neighborhoods', *American Journal of Community Psychology* 9(1): 55–66.
Robbins, E. (2004) 'New urbanism', in E. Robbins and R. El-Khoury (eds), *Shaping the City: Studies in history, theory and urban design*, London: Routledge, pp. 212–30.
Robertson, J. (1984) 'The empire strikes back,' In D. Porphyrios (ed.), *Leon Krier, Houses, Palaces, Cities*, London: Architectural Design Editions, pp. 11–3, 18–9.
Rogers, R. (2004) 'City revival on course', *The Guardian* Unlimited, 5 August 2004,

Online. Available at http://society.guardian.co.uk/urbandesign/story/0,11200, 1276130,00.html (Accessed 7 August 2004).
Roman Britain Organization (2003) 'Calleva Atrebatum', Online. Available at http://www.romanbritain.org/places/calleva.htm (Accessed 6 August 2004).
Ross, A. (1999) *The Celebration Chronicles: Life, liberty, and the pursuit of property value in Disney's new town*, New York: Ballantine Books.
Rowe, P. (1991) *Making a Middle Landscape*, Cambridge, MA: MIT Press.
Rubinstein, D. (1974) *Victorian Homes*, London: David and Charles.
Russell, J. (2001) 'A new new urbanism renews Dutch docklands', *Architectural Record* 189(4): 94–102.
Sahlins, M. (1976) *The Use and Abuse of Biology: An anthropological critique of sociobiology*, Ann Arbor: University of Michigan Press.
Salingaros, N. (2004) 'Leon Krier' webpage. Online. Available at http://applied.math.utsa.edu/krier/ (Accessed 1 August 2004).
Sanchez, T. W. and R. E. Lang (2002) *Security versus status: the two worlds of gated communities*, Draft Census Note 02:02 (November) Metropolitan Institute at Virginia Tech, Online. Available at http://www.mi.vt.edu/Files/Gated%20Census%20Note1119.pdf (Accessed 12 October 2003).
Sandercock, L. (1998) *Towards Cosmopolis: Planning for multicultural cities*, Chichester UK: John Wiley.
Sauer, L. (1994) 'Creating a "signature" town: The urban design of Bois-Franc', *Plan Canada* 34(4): 22–7.
Schaffer, D. (1982) *Garden Cities for America: The Radburn experience*, Philadelphia: Temple University Press.
Schaffer, F. (1970) *The New Town Story*, London: MacGibbon and Kee.
Scully, V. (1994) 'The architecture of community', in P. Katz (ed.), *The New Urbanism: Toward an architecture of community*, New York: McGraw-Hill, pp. 221–30.
—— (1991) 'Seaside and New Haven', in A. Krieger (ed.), *Andres Duany and Elizabeth Plater-Zyberk: Towns and Town-making Principles*, Harvard University Graduate School of Design. New York: Rizzoli, pp. 17–20.
Seaside Institute (2004) 'Seaside is born', Online. Available at http://theseasideinstitute.org/page.aspx?mode=p&s=86420.79.7801 (Accessed 28 September 2004).
Sewell, J. (1993) *The Shape of the City: Toronto struggles with modern planning*, Toronto: University of Toronto Press.
—— (1977) 'The suburbs', *City Magazine* 2(6): 19–55.
Shelton, B. (1999) *Learning from the Japanese City: West meets East in urban design*, London: E & FN Spon.
Shibley, R. (2002) 'Placemaking as a critique of new urbanism', *Planners Network* 151: 6–8.

Shipley, R. (2000) 'The origin and development of vision and visioning in planning', *International Planning Studies* 5(2): 225–36.

Shipley, R. and R. Newkirk (1998) 'Visioning: Did anyone see where it came from?' *Journal of Planning Literature* 12(4): 407–16.

Simmons, I. G. (1996) *Changing the Face of the Earth*, Oxford: Blackwell.

—— (1993) *Environmental History: A concise introduction*, Oxford: Blackwell.

Simpson, M. (1985) *Thomas Adams and the Modern Planning Movement: Britain, Canada and the United States, 1900–1940*, London: Alexandrine Press Book.

Sitowski, R. (2004) Presentation in workshop, American Planning Association conference, Washington DC, 25 April 2004.

Sitte, C. (1965 [1889]) *City Planning According to Artistic Principles* (trans. G. Collins and C. Crasemann Collins), New York: Random House (reprint).

Smart Growth Network (2004) 'About Smart Growth', Online. Available at http://www.smartgrowth.org/about/default.asp (Accessed 14 September 2004).

Smith, J. (2002) 'HOPE VI and the new urbanism: eliminating low-income housing to make mixed-income communities', *Planners Network* 151: 22–5 (also Online. Available at http://www.plannersnetwork.org/htm/pub/archives/152/smith.htm (Accessed 14 September 2004).

Smith, M. (2004) 'Building on our past', *The Guardian* Unlimited. 6 August. Online. Available at http://society.guardian.co.uk/urbandesign/comment/0,11200, 1277618,00.html (Accessed 7 August 2004).

Soja, E.W. (1989) *Postmodern Geographies: The reassertion of space in critical social theory*, London: Verso.

Song, Y. and G.-J. Knaap (2003) 'New urbanism and housing values: a disaggregate assessment', *Journal of Urban Economics* 54: 218–38.

Sorensen, A. (2004) 'Major issues of land management for sustainable urban regions in Japan', in A. Sorensen, P. Marcotullio, and J. Grant (eds), *Towards Sustainable Cities: East Asian, North American and European perspectives on managing urban regions*, Aldershot: Ashgate, pp. 197–216.

—— (2002) *The Making of Urban Japan: Cities and planning from Edo to the 21st century*, London: Routledge.

—— (1999) 'Land readjustment, urban planning, and urban sprawl in the Tokyo metropolitan area', *Urban Studies* 36(13): 2333–60.

Southworth, M. (1997) 'Walkable suburbs? An evaluation of neotraditional communities at the urban edge', *Journal of the American Planning Association* 63(1): 28–44.

Southworth, M. and E. Ben-Joseph (1995) 'Street standards and the shaping of suburbia', *Journal of the American Planning Association* 61(1): 65–81.

Southworth, M. and P. Owens (1993) 'The evolving metropolis: studies of community, neighborhood, and street form at the urban edge', *Journal of the American Planning Association* 59(3): 271–87.

Springer, J. (2000) 'A solution for sprawl? Federal's Street Retail unit paves the way', *Shopping Centers Today*, January. Online. Available at http://www.icsc.org/srch/sct/current/sct0100/01.html (Accessed 10 April 2004).

Staley, S. (1997) 'The new urbanism: an overview', Urban Futures Organization, Reason Public Policy Institute. Online. Available at http://www.urbanfutures.org/r6897a.html (Accessed 7 October 2004).

Statistics Canada (2004) 'Type of dwelling and population by type of dwelling (1961–2001 Censuses)', Online. Available at http://www.statcan.ca/english/Pgdb/famil66.htm (Accessed 16 October 2004).

Stead, D. and E. Hoppenbrouwer. (2004) 'Promoting an urban renaissance in England and the Netherlands', *Cities* 21(2): 119–36.

Steil, L. with N. Salingaros. (2004) 'Leon Krier- selected buildings', Online. Available at http://zakuski.utsa.edu/krier/BUILDINGS/krierbuildings.html (Accessed 6 August 2004).

Stein, R. (2003) 'Urban sprawl linked to obesity', *The Age* Online. Available at http://www.theage.com.au/articles/(2003)/08/29/1062050668503.html?one click=true (Accessed 14 August 2004).

Steiner, R. (1998) 'Traditional shopping centers', *Access* 12: 8–13.

Steuteville, R. (2004a) 'New urban neighborhoods make big gains', *New Urban News* 9(1): 1, 3–6.

—— (2004b) 'Numbers don't lie: HOPE VI has worked wonders', *New Urban News* 9(4): 2.

—— (2004c) 'New model proposed for waterfront town', *New Urban News* 9(4): 9–10.

—— (2003a) 'Britain looks to NU for major growth initiative', *New Urban News* 8(8): 1, 4.

—— (2003b) 'New urbanism does not promote crime', *Planetizen*, November 3. Online. Available at http://www.planetizen.com/oped/item.php?id=110 (Accessed 6 August 2004).

Strupat, B. (2001) 'The many why's of Windsor', *Vero Beach Magazine* (March/April) Online. Available at http://www.effective-writer.com/Author/windsor-article.htm (Accessed 19 June 2004).

Swope, C. (2001) 'Rehab refugees', *Governing Magazine* (May), Congressional Quarterly. Online. Available at http://www.governingmagazine.com/archive/2001/may/housing.txt (Accessed 9 June 2004).

Tae, H. H. (1962) *Korea – Forty Three Centuries*, Korean Cultural Series, Volume 1, Seoul: Yonsei University Press.

Tait, M. (2003) 'Urban villages as self-sufficient, integrated communities: a case study in London's Docklands', *Urban Design International* 8: 37–52.

Talen, E. (2003) 'Andres Duany on "Why write codes?"', PLANET, Online posting. Available email: PLANET@LISTSERV.BUFFALO.EDU (15 May 2003).

—— (2002) 'The social goals of new urbanism', *Housing Policy Debate* 13(1): 165–88.

Talen, E. (2001) 'Traditional urbanism meets residential affluence: an analysis of the variability of suburban preference', *Journal of the American Planning Association* 67(2): 199–216.
—— (2000a) 'The problem with community in planning', *Journal of Planning Literature* 15(2): 171–83.
—— (2000b) 'Consumer sovereignty and new urbanism', PLANET, Online posting. Available email: PLANET@LISTSERV.BUFFALO.EDU (28 September 2000).
—— (2000c) 'Measuring the public realm: a preliminary assessment of the link between public space and sense of community', *Journal of Architectural and Planning Research* 17(4): 344–60.
—— (1999) 'Sense of community and neighborhood form: an assessment of the social doctrine of new urbanism', *Urban Studies* 36(8): 1361–79.
Talen, E. and C. Ellis. (2002) 'Beyond relativism: reclaiming the search for good city form', *Journal of Planning Education and Research* 22: 36–49.
Teitz, M. (1998) 'New urbanism: fashion trend or movement?', *Town and Country Planning* 67(1): 41.
Thompson-Fawcett, M. (2003a) ' "Urbanist" lived experience: resident observations on life in Poundbury', *Urban Design International* 8: 67–84.
—— (2003b) 'A new urbanist diffusion network: the Americo-European connection', *Built Environment* 29(3): 253–70.
—— (2000) 'The contribution of urban villages to sustainable development', in K. Williams, E. Burton and M. Jenks (eds), *Achieving Sustainable Urban Form*, London: E & FN Spon. pp. 275–87.
—— (1998) 'Leon Krier and the organic revival within urban policy and practice', *Planning Perspectives* 13: 167–94.
Till, K. (1993) 'Neotraditional towns and urban villages: the cultural production of a geography of "otherness" ', *Environment and Planning D: Society and Space* 11: 709–32.
Tirinnarri, G. (2004) 'Taking new urbanism to a new level', Presentation at American Planning Association conference, Washington DC, 25 April 2004.
Town, S. (2003) 'New urbanist article', Planetizen, November 20. Online. Available at http://www.planetizen.com/oped/cmt_item.php?id=1312 (Accessed 6 August 2004).
Tu, C. C. and M. J. Eppli (1999) 'Valuing new urbanism: the case of Kentlands', *Real Estate Economics* 27(3): 425–51.
Turan, B. (2004) 'Architecture and technê: the impossible project of Tendenza', *Architronic* 7(1) Online. Available at http://architronic.saed.kent.edu/v7n1/v7n104a.html (Accessed 12 September 2004).
Turner, F. (1991) *Beauty: The value of values*, Charlottesville VA: University Press of Virginia.

UK National Statistics (2002) 'Cars or vans: 1972 to 2002', Online. Available at http://www.statistics.gov.uk/STATBASE/ssdataset.asp?vlnk=8066&More=Y (Accessed 5 August 2004).
University of Westminster. (2004) 'Gated communities: good for social inclusion and urban renewal' (Press release, June 22), Online. Available at http://www.westminster.ac.uk/page2903 (Accessed 30 August 2004).
Unwin, R. (1912) *Nothing Gained by Overcrowding*, Pamphlet. London: Garden Cities and Town Planning Association.
Upton, R. (2002) 'Planning praxis: ethics, values and theory', *Town Planning Review* 73(3): 253–69.
Urban Task Force (1999) *Towards an Urban Renaissance*, Final Report of the Urban Task Force under Lord Rogers of Riverside, Department of the Environment, Transport and the Regions, London: E & FN Spon.
US Census (2002). 'Demographic trends in the 20th century', Online. Available at http://www.census.gov/prod/2002pubs/censr4.pdf (Accessed 16 December 2004).
Vancouver, City of (n.d.) 'Southeast False Creek: a sustainable urban neighbourhood and major park on the False Creek Waterfront' (pamphlet), Vancouver: Planning Department.
Van der Ryn, S. and P. Calthorpe. (1986) *Sustainable Communities: A new design synthesis for cities, suburbs and towns*, San Francisco: Sierra Club Books.
Van Tilburg, J. (2000) 'Disguising density', *Urban Land* 59(7): 58–9.
Veregge, N. (1997) 'Traditional environments and the new urbanism: a regional and historical critique', *TDSR* VIII(II): 49–62.
Vidler, A. (1968) 'The idea of unity and Le Corbusier's urban form', in D. Lewis (ed.), *Urban Structure*, Architects' Yearbook XII, London: Elek Books, pp. 225–237.
Vischer, J. (1984) 'Community and privacy: planners' intentions and residents' reactions', *Plan Canada* 23(4): 112–22.
Wackernagel, M. and W. Rees (1995) *Our Ecological Footprint: Reducing human impact on the Earth*, Gabriola Island, BC: New Society Publishers.
Ward, S. (1998) *Selling Places: The marketing and promotion of towns and cities 1850–2000*, London: E & FN Spon.
Ward, S. (ed.) (1992) *The Garden City: Past, present, and future*, London: E & FN Spon.
Warson, A. (1996) 'Toronto builder commits to Duany plan', *Planning* 62(12): 21.
Watanabe, S.-I. (1992) 'The Japanese garden city', in S. Ward (ed.), *The Garden City: Past, present and future*, London: E & FN Spon, pp. 69–87.
Waterloo, City of (1998) *West Side Nodes Zoning Study Discussion Paper*, Waterloo ON: Development Services.
Webster, C. (2002) 'Property rights and the public realm: gates, greenbelts, and Gemeinschaft', *Environment and Planning B: Planning and Design* 29(3): 397–412.

Webster, C., G. Glasze and K. Frantz (2002) 'The global spread of gated communities', *Environment and Planning B: Planning and Design* 29(3): 315–20.

Webster, C. and L. W.-C. Lai (2003) *Property Rights, Planning and Markets: Managing spontaneous cities*, Cheltenham, UK: Edward Elgar.

Wegener, M. (1994) 'Tokyo's land market and its impact on housing and urban life', in P. Shapira, I. Masser and D. Edgington (eds), *Planning for Cities and Regions in Japan*, Liverpool: University of Liverpool Press, pp. 92–112.

Wexler, H. (2001) 'HOPE VI: market means / public ends. The goals, strategies and midterm lessons of HUD's urban revitalization demonstration program', *Journal of Affordable Housing* 10(3): 195–233.

White, R. (1996) 'Sustainable urban communities: Calgary's approach', *Plan Canada* 36(4): 16–9.

Wiegandt, C.-C. (2004) 'Mixed land use in Germany: Chances, benefits and constraints', Paper prepared for the invitational workshop, 'Incentives, regulations and plans: The role of states and nation states in smart growth planning', Annapolis, Maryland: National Center for Smart Growth Research and Education. Online. Available online at http://www.smartgrowth.umd.edu/InternationalConference/ConferencePapers/Wiegandt_Mixed per cent20Use.pdf (Accessed 2 November 2004).

Wight, I. (1996) 'In search of grander, humane visions', *Plan Canada* 36(4): 3–4.

—— (1995) 'New urbanism *vs* conventional suburbanism', *Plan Canada* 35(5): 20–2.

Will, G. (2004) 'Waging war on Wal-Mart', *Newsweek* 5 July: 64.

Williams, K., E. Burton and M. Jenks (eds) (2000) *Achieving Sustainable Urban Form*, London: E & FN Spon.

Windsor, Florida (2004) 'Windsor', Web page. Online. Available at http://www.windsorflorida.com/facts.html (Accessed 15 September 2004).

Winnipeg, City of (2001) *Plan Winnipeg 2020 Vision*, Winnipeg, Manitoba, Canada.

Witty, D. (2002) 'Healthy communities: what have we learned?', *Plan Canada* 42(4): 9–10.

Wood, D. (2003) 'Editorial: Foucault and panopticism revisited', *Surveillance and Society* 1(3): 234–9. Online. Available at http://www.surveillance-and-society.org/articles 1(3)/editorial.pdf (Accessed 29 May 2004).

World Commission on the Environment and Development (WCED) (1987) *Our Common Future*, Report of the Brundtland Commission. New York: Oxford University Press.

Wright, A. (1967) 'Chang'an', in A. Toynbee (ed.), *Cities of Destiny*. London: Thames and Hudson, pp. 138–49.

Wright, E. O. (1997) *Class Counts: Comparative studies in class analysis*, Cambridge UK: Cambridge University Press, Maison des Sciences de L'Homme.

Wu Liangyong (1986) *A Brief History of Ancient Chinese City Planning*, Urbs et regio. Kasseler Schriften zur Geographie und Planung, Kassel: Gesamthochschulbibliothek.

Wyatt, R. (2004) 'The great divide: differences in style between architects and urban planners', *Journal of Architectural and Planning Research* 21(1): 38–54.

Yamaguchi, E. (1999) 'Why it's impossible to build apartments to last 100 years', *Shukan Asahi* 21 May 1999, pp. 150–1.

Yamasa Institute (2004) 'Hikone City', Hattori Foundation, Aichi Prefecture. Online. Available at http://www.yamasa.org/japan/english/destinations/shiga/hikone.html (Accessed 24 August 2004).

York, Region of (1994a) *Cornell: Official Plan Amendment and Secondary Plan*, Region of York: Planning and Development Services.

—— (1994b) *Official Plan Amendment and Secondary Plan: Angus Glen*, Region of York: Planning and Development Services.

York, G. (2004) 'How could they demolish them so easily?', *Globe and Mail* (Toronto) GlobeandMail.com, 3 January 2004. Online. Available at http://www.the globeandmail.com/servlet/story/RTGAM.(2004)0102.wchina03/BNStory/ International/ (Accessed 8 January 2004).

Zelenak, E. (2000) 'Brave new urbanism', *Community Links* VII(2): Issue 11, Winter. Online. Available at http://www.communitypolicing.org/publications/comlinks (Accessed 12 August 2004).

Zetter, J. (1994) 'Challenges for Japanese urban policy', in P. Shapira, I. Masser and D. Edgington (eds), *Planning for Cities and Regions in Japan*, Liverpool: University of Liverpool Press, pp. 25–32.

Zimmerman, J. (2001) 'The "nature" of urbanism on the new urbanist frontier: sustainable development, or defense of the suburban dream?', *Urban Geography* 22(3): 249–67.

INDEX

accessibility 63, 97, 98, 122, 163, 176
Adams, Thomas 111
affordable housing xvii, 27, 37, 41, 69, 83, 86, 89, 94–6, 99, 110, 118–19, 125, 132, 135, 139, 155, 160, 162, 168, 170–1, 180–1, 188–90, 206, 213, 222, 226
Ahwahnee principles 56
Alessandria, Italy 127
Alexander, Christopher 10, 52, 219
Alfandre, Joe 85, 87
Alimanestianu, Joanna 112, 115
American Planning Association 65, 89, 96–8, 103, 197, 211, 220
Angus Glen (Markham ON) 161
Aqua (Florida) 99
Architects Guild 97
authenticity 53, 168, 181–3, 216
A Vision of Europe (AVOE) 113

Barcelona (Spain) 92, 105, 108, 176
Beach, The (Toronto ON) 165
Beasley, Larry 166
Beaux Arts 36, 83, 139, 217, 228
Beazley, Sue (contributor, Box 4.2) 94
Berczy Village (Markham ON) 161
Bethesda Row (Bethesda MD) 92
Beverly Hills CA 94
Bilbao (Spain) 108
Birmingham (UK) 108, 120
Blair, Tony 22
Bohl, Charles 175
Bois Franc (Montreal QC) 166
Bordesley (Birmingham) 120
Borneo Sporenburg (Amsterdam) 116
Bournville 34
Bruges (Belgium) 106–7
Burnham, Daniel 36
Bush (George W) administration 95
bye-law street 35

Cadbury, George 34
Calgary AB 155, 157–60, 167
Calthorpe, Peter 7, 17, 46–7, 53–4, 56, 59–60, 63, 82, 199; in Canada 157, 167; Laguna West 84

Calthorpe Associates 211, 217
Canada Lands Company 164–5
Canada / Central Mortgage and Housing Corporation (CMHC) 152, 161
Canadian Institute of Planners 65, 155
Canadian Public Health Association 155
Canadian Urban Institute 157
Canberra (Australia) 36
Carma Developments 157, 159–60
car use 109, 118, 123–4, 126, 131, 134, 139, 141, 164, 168, 178, 191, 207, 211, 215, 218
castle towns (Japan) 137, 147
CCTV– *see video surveillance*
Celebration FL 81–2, 88–9, 98, 100, 181, 186, 190, 193–4, 196, 204–5, 213, 219, 232ch5n2
Chang'an (Xian, China) 133, 135, 136
charrette 72, 76, 183–4, 185
Charter of European Urbanism 119
Charter of the New Urbanism 96, 166
Chiarmonti, Ray 213
China 132–4, 178
Chomsky, Noam 22
city beautiful movement 35–6, 38, 51, 217, 228
City Planning Areas (Japan) 145
civility 21–3, 50, 52, 100, 177, 195, 226
Clair Creek Village (Waterloo ON) 164, 182
class issues 96, 110, 127, 187, 190, 196–7
Cleveland OH 36
Clinton administration 95
codes xvii, 18, 22, 56, 65, 72, 82, 118–19, 168, 178, 186, 188, 189, 192, 194, 200, 213; as civilizing 195–7; form-based 65, 97, 103, 161, 197, 220; national building 160; Seaside 83, 220; SmartCode 65, 74, 97, 220; Windsor 84
collaborative planning 183, 201, 224
Columbia MD 214
commercial uses (viability of) 88, 90, 98–9, 115, 117, 120, 122, 124, 126–7, 140, 143–4, 159–60, 162, 182, 187, 208–9, 223; required 161

commodity consumption 12–13, 43, 69, 92, 100, 108, 144, 172, 184, 186, 193, 197, 200, 215, 216, 233ch8n1
common good – *see public interest*
communicative planning theory 15, 224
compact city theory 13, 112–3, 125, 133
concentric zone theory 75
Confusianism 132–3, 135, 137
Congress for the New Urbanism 3, 55, 61–2, 72, 81, 88, 96–8, 103, 156, 171, 180, 194, 206, 211, 220, 232ch4n1
Constantinople (Istanbul) 106
Corbusier – *see Le Corbusier*
Cornell (Markham ON) 151, 161–4
Correa, Jaime 81
Council for European Urbanism 113, 119
Crown Street (Glasgow) 124
Culot, Maurice 52, 112

danchi (Japanese suburban new town) 142
Davis, Robert 54, 82, 96
democratic participation 183–6, 199, 213, 222
design guidelines 45, 97, 118, 161, 163, 167, 172
Dewey, John 15
Disney 82, 88–9, 100, 193
diversity 12, 27, 70–1, 77, 100, 123, 129, 157, 177, 182–3, 185, 186–8, 193, 198, 200, 210, 215, 224, 226
Don Mills (Toronto) 152–3
Dover, Victor 81, 112
DPZ (Duany Plater-Zyberk & Company) 9, 57, 65, 83, 84, 97, 123, 211, 217, 220
Dresden (Germany) 107
Duany, Andres 7, 8, 17, 52–8, 63, 81–3, 85–8, 100, 112, 151, 185, 199, 211, 227; in Canada 157, 161, 163, 167
Dublin (Ireland) 108

Eastbridge (Waterloo ON) 164
East Lake (Atlanta GA) 95
East Ocean View (Norfolk VA) 100
ecological theory 16, 19–20, 63, 73–5, 191–2; landscape ecology 46
Edo era 137, 147
environmental concerns 100–1, 109, 120, 127–8, 131–2, 134, 141, 145, 149, 156, 161, 183, 190
environmental determinism – *see spatial determinism*
Environmental Protection Agency 97
Erskine, Ralph 120
Eurocouncil 112, 119

evolutionary theory 73, 132
expertise 18–19, 76–7, 185, 199, 200, 214, 224
Exposition / World's Fair Columbian (Chicago) 34, 36; Prague 34; Vancouver 166
eyes on the street 22, 52, 66, 126, 137, 148, 196

factory town 31, 34–5, 208
Fallsgrove (Rockville MD) 5, 90, 98
False Creek (Vancouver BC) 155, 166–7
Federal Realty 92, 94
Federation of Canadian Municipalities 155
Fedora 228
Fletcher Farr Ayotte 91
Fonti di Matilde (San Bartolomeo, Italy) 116
Forester, John 15, 201
Foucault, Michel 15, 22, 196
Foundation for Urban Renewal 113
Fourier, Charles 31
Friedman, Avi 166
Friedmann, John 201
friendly fascism 197

Garden Cities Association 38
garden city xix, 7, 14, 19, 29–44, 51, 65, 111, 128, 142, 152, 156, 171–2, 177, 203, 205, 208, 211–12, 217, 228–9; garden suburb 39–40, 47, 165; in Japan 40
Garrison Woods (Calgary AB) 164–5
gated developments xvii, 68, 83–4, 102, 126, 133, 134, 181, 203, 211, 216–7, 227–9
Gaudi, Antoni 108
Gehry, Frank 108
gentrification 16, 70, 92, 96, 99–100, 165, 178, 189, 214
Ghent (Belgium) 106
Glasgow (Scotland, UK) 107, 124
Gothic revival 110, 203
Greeks (classical) 9, 36, 105–6, 228
Green Plan (Canada) 156
Greenwich Millennium Village (London) 61, 120, 123, 128
Greenwich Village (New York) 14

Hampstead Garden Suburb (London) 39
Hampstead Norris (Norreys) UK 110
Hancock, Macklin 153
Hangzhou (Hangchow), China 133
Harbor Town (Memphis TN) 99
Healey, Patsy 201

healthy communities movement 55, 60, 124, 155–6, 219
Hikone, Japan 147
homeowner associations 89, 185–6
HOPE VI 71, 95–6, 99, 165, 180, 184
house size 102, 128, 135, 141, 169, 190
Howard, Ebenezer 36–9, 41–3, 47, 66, 111, 219
HUD (Housing and Urban Development) 82, 95
human nature xvi, 75, 77
hutong (traditional housing in China) 134

Imperial Hotel (Tokyo) 139
Institute of Transportation Engineers 97
INTBAU 113
Inverness Village (McKenzie Towne, Calgary) 159–60
I'On (Mount Pleasant SC) 99
Itoh Kaori (contributor, Box 6.1) 144

Jacobs, Jane 9, 14, 17, 51–2, 153–5, 219
just society 22, 76

Karow Nord (Berlin) 116–9
Katz, Peter 54, 62
Kelbaugh, Doug 7
Kentlands, The (Gaithersburg MD) 58, 84–8, 98–9, 100, 160, 162, 181, 219
King Farm (Rockville MD) 85, 89–91; shuttle bus 89
Kings, The (Toronto) 165
Kohl, Joe 81
Korea, South 135–6, 185
Krier, Leon 3, 7, 9, 14, 52, 54–5, 63, 73, 81, 111–12, 120, 129, 199, 227
Krier, Rob 52, 124
Krier Kohl 211, 218
Kropotkin, Peter 37
Kunstler, James Howard 54, 68, 100, 211
Kurokawa, Kisho 142
Kyoto (formerly Heiankyo, Japan) 136, 137

Laguna West CA 84
Lakelands (Gaithersburg MD) 208
Las Ramblas (Barcelona) 92
Las Vegas NV 94
Lasserre, Christian 115
Law Developments 163
Le Corbusier 40, 139
Lennertz, William 81
Letchworth Garden City (UK) 38
Lever, William 34
Linzi (China) 133

Lisbon (Portugal) 107
Liverpool (UK) 108
London 37–8, 41, 61, 105–6, 108, 109, 120–1
lot size 101–2, 119, 124, 128, 135, 137, 152, 156, 158, 159, 167, 190–1, 215
Louis Napoleon (Napoleon III) 108, 194
Louis XIV 194
Lynch, Kevin 10–11, 14, 23, 52

machizukuri (municipal ordinances in Japan) 145, 185
Manchester (UK) 107
Markham ON planning policy 157, 161–4, 169, 190
Markham Town Centre 161, 170
Marseilles (France) 106
McHarg, Ian 219
McKenzie Lake (Calgary) 158, 160
McKenzie Towne (Calgary) 157–60, 162, 181
Melrose Arch (South Africa) xvii, 184
millennium villages 60; see *also* Greenwich Millennium Village
Moan, Patrick (contributor, Box 4.1) 86
modernism 6, 11, 18–19, 40–2, 44, 47, 51, 77, 107, 110–12, 134, 136, 139, 145–6, 149, 177, 195, 200, 217–18, 224, 228; modernist new urbanism 166, 176
Montreal QC 155, 165–6
Moore Ruble Yudell 117
Moule, Elizabeth 56
Mumford, Lewis 106, 111
Murasaki Shikibu 136

Nagoya (Japan) 138, 139, 141, 142
Napoleon III, see *Louis Napoleon*
Nara (formerly Heijokyo, Japan) 136
National Town Builders Association 97
nature versus culture 45–6, 56, 74
neighborhood concept 40, 41, 62, 65, 99, 162, 227, 232ch3n2
New Delhi (India) 36
New Harmony IN 31
New Lanark (Scotland) 31
New York NY 92, 132
Niagara-on-the-Lake ON 164
Nisshin (Japan) 141
Nordnes (Norway) 47–9
Nordquist, John 113
normative approaches and theory 16, 24, 73–4, 200, 219, 220, 223

Oakville ON 164
Office of the Deputy Prime Minister (UK) 65, 113
Orangeville ON 164
Orenco Gardens (Hillsboro OR) 91
Orenco Station (Hillsboro OR) 59, 91, 180
organic analogy 19, 73, 192
Osaka (Japan) 139, 141
Ouellet, Marc (contributor, Box 5.2) 123
Owen, Robert 31, 37

Pacific Realty Trust 91
panopticon 196
Paris (France) 105-6, 109, 194
Parker, Barry 38-41
Parsons, Talcott 75
pattern book 88, 97, 220
Pearl District (Portland OR) 92
Penrose Group 89
Pentagon Row 92
Perry, Clarence 40-2, 99, 219
planner, role of 145, 220-3
planning guidance notes or policy 113, 114, 164, 171, 220
Plater-Zyberk, Elizabeth 7, 52, 54-8, 81-3, 85-8, 112
Pleasantville 167
Plum Creek (Kyle TX) 102
phalanstery 31
Phoenix Trust 112, 113
Pitsford (Northampton UK) 109
Polo, Marco 133-4
Polyzoides, Stefanos 56, 112
Portland OR 91-2
Portland Hills (Halifax NS) 204-5
Port Sunlight (UK 34
post-modernism 46, 77
Poundbury (Dorset UK) 58, 65, 112, 120-3, 125-6, 127, 181, 194, 197, 219
power 15, 73, 76-8, 192-9, 201, 203, 206, 212, 223, 225
Prairie Crossing IL 190, 232ch3n4
Prescott, John 113
Prince of Wales, Charles 4, 23, 53-4, 60, 111-12, 120, 129, 194
Prince's Foundation 65, 112, 113
Prince Regent 108
private communities 171, 185-6, 189, 196, 197, 214
private streets 163, 194, 214
Providence RI 100
public housing xvii, 27, 40, 44, 69, 71, 94-6, 110, 118, 120, 126, 165, 178, 180, 184, 188, 215

public interest xvi, 11, 19, 67, 71, 200, 219, 220, 224
public realm 52-3, 56, 63, 72, 82, 94, 131, 138, 142, 146, 148, 166, 170, 171, 176, 181, 189, 214
Pullman IL 34
Pyongyang (Korea) 135

Queen Anne style 38, 110, 203
quarter (quartier) 52, 73, 82, 111, 227

racial or ethnic issues 86, 91, 110, 194, 215
Radburn NJ 42, 142, 153
rational planning 77, 219, 224
Reagan, Ronald 81
Regent Park (Toronto) 165
regional planning 59, 72, 151, 184, 201, 222, 233ch7n1
Regional Planning Association 41
regulatory reform 35
residents associations - *see homeowner associations*
retrofitting suburbs 92
Richmond, Cora 37
Robertson, Jaquelin 83, 88
Rogers, Lord Richard 112
Romans (classical) 9, 36, 106, 127, 228
Rome (Italy) 107, 108, 194
Rosemary Beach (Walton County, FL) 99
Rossi, Aldo 52
Round, The (Beaverton OR) 92
Royal Institute of British Architects (RIBA) 3, 128-9
Rue de Laeken (Brussels, Belgium) 115, 124, 127

St Augustine FL 194
St Lawrence (Toronto) 154
St Joe Company 83
Santana Row (San Jose CA) 92-4, 99
Sarasota FL 97
Saunders, Todd (contributor, Box 3.1) 49
Savannah GA 166
Sears, Trevor (contributor Box 5.1) 119
Seaside FL 4, 9, 54-5, 58, 65, 82-4, 85, 99, 151, 219, 231n4
Seaside Institute 82, 96
Seaton competition 161
Section 8 rent vouchers 95-6
security xvii, 66, 68, 98, 105, 126, 177, 216-17, 227-9
segregation 91, 110, 136, 187-8
sense of community 69, 71, 88, 120, 162, 167, 171, 178, 205, 207, 209-10

sense of place 65, 71, 83, 178, 207–8, 226
Seoul (Korea) 132, 135–6
Sewell, John 4, 154–5
Sitte, Camillo 36, 51, 110, 111, 219
skinny house 169
social housing (*see public housing*)
social mix and integration 70, 96, 119, 123, 127, 170, 189, 198, 206, 226
sociobiology 75
SongBrook (Eugene OR) 204–5
spatial determinism (and environmental affordance) 6, 27, 67–8, 71, 175–6, 210
spatial ethics 15, 194, 225
Steil, Lucien 112
Stern, Robert 88
Stevenson, J.J. 203
strategic planning 155, 183
street patterns 106, 108, 126, 176–7; cul-de-sac 84, 126, 144, 152, 161, 170; grid layout 26, 51, 58, 63, 66, 88, 93, 105, 124, 136, 137, 142–3, 162, 164; grid in Asia 133, 135; street naming 146, 148
Stuart FL 185
style 71–2, 101, 216; theme park 92, 147
surveillance 22, 66, 100, 148, 196
sustainable development 55–6, 110, 112–13, 124–5, 127–8, 144, 155–6, 161, 169, 190–2, 219

Tagliaventi, Gabriele 112
Tagliaventi and Associates 218
Talen, Emily 7, 63, 175
Tale of Genji, The 136
Tokugawa Shogunate 137
Tokyo (formerly Edo, Japan) 132, 140, 141, 231ch1n7
Toronto ON 36, 151, 153–5, 167
tourism 110, 122, 146, 147, 208–9
traffic jams, gridlock 133, 136, 141, 144, 211
transect 19, 63, 74–5, 97, 101, 163
travel behaviour 100, 141; commuting time 108, 141
Truman Show, The 4, 82

universal principles 10–12, 52, 58, 76, 111, 176–7, 198, 213–14, 219, 224–5, 227

universities linked to new urbanism: Catolica Portuguesa 114; Ferrara 114; Harvard 211; Miami 81, 211; Notre Dame 114, 211
Unwin, Raymond 38–41, 51, 111, 219
Upton Village (Northampton UK) 115, 125
Urban Land Institute (ULI) 65, 82, 96–8, 103, 211
Urbanization Control Areas (Japan) 145
Urbanization Promotion Areas (Japan) 145
urban renaissance 3, 8, 61, 112–15, 124–5, 128–9, 222
urban renewal 11, 44, 61, 152, 153, 155, 165
Urban Strategies 211
Urban Task Force 112, 128
Urban Villages Group/Forum 55, 60–1, 112, 194
utopian commune 29, 31–4

Vancouver BC 36, 154–5, 166–7, 180
Venice (Italy) 105–6
Versailles (Paris) 194
video surveillance 22, 126
visioning and visions 24–5, 167, 183, 185, 196, 200, 204, 218, 221
visitability – *see accessibility*

Washington DC 36, 85–6, 89–92
Waterloo ON 164
WaterColor FL 83
Weston, Galen 4, 54, 61, 84, 194, 232ch4n1
West Silvertown (London) 126–7
White City (Chicago) 36
White Town (Tajimi, Japan) 142–4
Wimpey Developers 114, 125
Windsor ON 164
Windsor (Vero Beach, FL) 84, 216
Woodlands, The (Texas) 214
World's fair / *see exposition*
Wright, Frank Lloyd 139

York UK 106
York Region ON 161, 169–70

zoning 3, 6, 18, 42, 50, 64–5, 85, 95, 97, 111, 124, 151, 157, 197; Euclidean 6

eBooks – at www.eBookstore.tandf.co.uk

A library at your fingertips!

eBooks are electronic versions of printed books. You can store them on your PC/laptop or browse them online.

They have advantages for anyone needing rapid access to a wide variety of published, copyright information.

eBooks can help your research by enabling you to bookmark chapters, annotate text and use instant searches to find specific words or phrases. Several eBook files would fit on even a small laptop or PDA.

NEW: Save money by eSubscribing: cheap, online access to any eBook for as long as you need it.

Annual subscription packages

We now offer special low-cost bulk subscriptions to packages of eBooks in certain subject areas. These are available to libraries or to individuals.

For more information please contact webmaster.ebooks@tandf.co.uk

We're continually developing the eBook concept, so keep up to date by visiting the website.

www.eBookstore.tandf.co.uk